AFRICA WHY?

(Lessons from nature to chart a path forward)

By

ELVIS MULUH

BOOK SOCIETY
OF NIGERIA

AF797597

Africa Why? By Elvis Muluh

First published in June 2022 by Book Society of Nigeria/ Magnet Africa Ltd. Printed and bounded in Abuja, Nigeria. Visit http:// www.magnet-africa.com, http://www.book-society.com

Published by Amazon Publishing in November 2022

TABLE OF CONTENT

Title i

Acknowledgment ii

Introduction..v

Dedication...viii

CHAPTER 1
Africa 14

CHAPTER 2
Manifesto for Colonialists 26

CHAPTER 3
The Great Deception 38

CHAPTER 4
The Story Untold..51

CHAPTER 5
The Arrival of the Scavengers.................................65

CHAPTER 6
Religion..78

CHAPTER 7
Ghost of low self-esteem..94

CHAPTER 8
The Unpleasant Place...108

CHAPTER 9
HEEDLESS LEADERS..120

CHAPTER 10
The Land of Vanished Dreams......................................133

CHAPTER 11
The Unlearned Lesson..144

CHAPTER 12
The Vital Role of "Repatriates"......................................162

CHAPTER 13
Not all Buffaloes are foolish after all 183

INTRODUCTION

"Imagine Africa where humans
learn lessons from the animal kingdom."

A pride of about seven lions lays an ambush against an advancing herd of buffaloes, they are in their hundreds. The craft and intelligence displayed by these big cats have always left me awestruck. I struggle to relax as I watch the unfolding scene with bated breath. Such documentaries on the National Geographic channel, juxtaposing intelligence and foolishness always keep me at the edge of my seat. I try hard to think of what is going on in the minds of these few lions; perhaps this pride has gone for days without food, and are desperate for a catch, or feeling cocksure and eager to show it. The odds look stacked against them but I tacitly wish them luck in their attack.

The herd approaches slowly, almost majestically, oblivious to the perils in their trail. My body tenses up some more as the camera pans to the pride. The lioness in charge of the ambush crouches in the swaying savannah grass. The others follow suit in different positions. I hold my breath and watch in curiosity, with hope that she executes her mission flawlessly. Time freezes and I begin to imagine; "If I were to be an animal or even possess a totem, it would certainly be a big cat, one of these apex predators which dominate the jungle. I'll be a ferocious one whose mere approach and roar as it nears the waterhole is capable of sending even the largest of beasts in the wild on their heels running for safety, leaving me in quietness and serenity to drink to my fill."

"Why were some animals created to feed on others? It's not fair at all" I ponder, shaking my head in perplexity.

The answers that pop into my mind are still wanting. Yes, I've followed reports about food chains and the ecosystem, how some animals are wired to feed on grass, while others feed on other animals to keep the ecosystem stable.

Oftentimes, I pity the helpless animals that fall prey to merciless predators. I have been frustrated at times with useable producers who continue to film unmoved while the prey groans in agony. I imagine the thoughts going on in the dying animal's mind as it fights to hold on to dear life. The cruelty becomes unbearable when the animal is a nursing mother whose baby is targeted. The mother puts up a resistance but before too long, it succumbs to the aggressors and watches helplessly, her baby being devoured while steering clear of the more savage fellows.

Back to my little screen, the buffaloes are now closing in. The leading lioness seemingly picks her prey then lays back getting ready to pounce, but she must wait for the others to stir the herd of buffaloes into a commotion. They do so when they realize the buffaloes are within eyeshot. As the lions emerge from beneath the bushes, the herd takes off in all directions. The pride of lions has their target in mind - the most vulnerable, they isolate the targeted buffalo and steer it towards the lioness that is ready to strike. The lioness launches from where it had taken cover, sinking its claws into the helpless prey, aiming at its throat, strangling it. As it shrieks and wriggles, vibrant life force is choked out. They both tumble over, in a twinkle of an eye the pride of lions joins her on the catch, each one focusing on devouring different body parts of the prey.

While I watch the lions feast on their catch, I can't help but wonder where such intelligence comes from; isolating an animal from amongst a group tens of times bigger and stronger than yours and strangling it.

"What if these animals only knew they were not as vulnerable as they are?" The thought keeps ringing in my mind. "With their large numbers and physical prowess, they would easily overpower the lions, trample on them given that the buffaloes outnumbered the lions about fourteen to one. If only they knew what agreeing as one means" I mutter as the lions devour their meal.

Unfortunately for the buffaloes, strength in numbers comes without the intelligence to corroborate. After completing their meal, the pride engages in some play, rolling over on the grass and enjoying each other. Once more, my thoughts return to the buffalo which was just consumed as a meal, "What if that were a nursing mother?" It saddens me to think that the baby might be lonely right now or even starving and wondering. My conclusion is simple: In this world of theirs, strength is important, but intelligence is supreme!

Watching animals in their natural habitat, with different types of skills and instincts for survival tactics is a marvel. Some are small in size, yet possess great ability to protect themselves through cooperation. Others, though bigger and stronger than all others in the jungle, turn out to be the most vulnerable, yet none lives forever because the ecosystem depends on them all.

DEDICATION

I dedicate this book first and foremost to my parents, Ndi Thomas Muluh and Ndi Ngwe Justin, who never had the opportunity to get an education themselves but against all odds that my siblings and I stopped at nothing to obtain one. "This poverty, nothing but education will overcome", Dad always told us in order to encourage us not to relent in education.

To my siblings, particularly Hilda Bih Muluh who instilled in me the essence of writing and emboldened me to carry out this project from the beginning till completion. I dedicate this book to you all (Walters Muluh, Hilda Muluh, Honorine Muluh, Britha Muluh, Eugene Muluh, Mabel Muluh, Javis Muluh and Kluivert Muluh and Linda Fan)

To my wife, Claudine Forche; my beautiful children, Nsudoh Muluh and Asa'ah Muluh. The uncountable times that I have taken away from you in order to see this dream become a reality, I dedicate this book to you with much love.

To all my teachers from C.S Alahkuma where I started my journey of acquiring an education through G.S Buchi Central and G.S.S Kwang Pinyin to B.G.S. Molyko Buea, I thank you all especially Mr & Mrs Ayimgeh Paul who were not only my teachers but parents to me as well. I remain indebted to you all and therefore dedicate this book to you.

To the fearless patriots of Africa whose words and works I have combined to write this book, it is a dedication to the sacrifices that even cost your lives. (Ruben Um Nyobe, Nnamdi Azikiwe, Tafawa Balewa, Kwame Nkrumah, Jerry Rawlings, Ahmed Sekou Toure, Leopold Seda Senghor, Ahmed Ben Bella, Jomo Kenyata, Julius Nyerere, Patrice Lumumba, Samora Machel, Emperor Haile Selassie, Thomas Sankara, Steve Biko, Nelson Mandela, Muammar al-Gaddafi, Zenzile Miriam Makeba, Bob Marley, Marcus Garvey, W.E.B. Du Bois, Rev. M.L. King Jr.) amongst the countless sons and daughters of mother Africa who left an indelible mark that will live forever. I salute your sacrifices to give us descendants of Africa a better place in this jungle called earth.

To the many tireless seekers of truth and adventure who do not cease to bring us the marvel of the animal world and injustices. I dedicate this book to you all as well.

To my publishers, thank for your relentless work from the very first day. You all have shown what professionalism entails. I dedicate this book to you all.

To the creator of this jungle, I dedicate this book to God Almight.

CHAPTER 1

AFRICA

Beginning of an endless exploitation

CHAPTER 1

AFRICA

Beginning of an endless exploitation

The geography of North Africa has been reasonably well known among Europeans since classical antiquity in Greco-Roman geography.

European exploration of Sub-Saharan Africa begins with the "age of discovery" in the 15th century, pioneered by the kingdom of Portugal under Henry the Navigator. The Cape of Good Hope was first reached by Bartolomeu Dias on 12 March 1488, opening the important sea route to India and the Far East, but European exploration of Africa itself remained very limited during the 16th and 17th centuries. Contemporary European powers were content to establish trading posts along the coast, while actively exploring and colonizing the New World.

At the beginning of the 19th century, European knowledge of the geography of the interior of Sub-Saharan Africa was still rather limited. Expeditions exploring Southern Africa were made during the 1830s and 1840s, around the midpoint of the 19th century and the beginning of the colonial scramble for Africa. The unexplored parts were now limited to what would turn out to be the Congo Basin and the African Great Lakes. Amongst the famous explorers who opened up the interior of Africa were John Hanning Speke, Sir Richard Burton, David Livingston and Henry Morton Stanley as stated in Wikipedia[1]

"The Portuguese explorers led by Fernando Pô are known to be the first Europeans to arrive off the coast of Cameroon in 1472. They reached Douala through the estuary of the Wouri River. They named it "Rio dos Camarões" (River of Prawns), whose anglicized version became Cameroon. With the arrival of the Europeans, business ensued between the locals and the foreigners. Slave trade was one of the businesses that quickly developed between them. Local Chiefs, especially those at the coast grew more powerful by signing agreements with the Portuguese. As time went on, business was also negotiated with traders from other European countries like England, Holland, France and Germany.

The local chiefs served as middlemen between the Europeans and local tribesmen. Mostly slaves, ivory and timber were exported from Cameroon as well as other countries where the Europeans were present. The Europeans sold clothes and metal-products to the locals. This perspective was first brought to my attention in primary (class) seven as our teacher, Mme Diana Achidi taught us Cameroon's history. The coming of the Whiteman led to the introduction of products such as whiskey, mirrors, and guns amongst many others. With the coming of these explorers, a doorway was opened to the beauty, diversity and rich potential of Africa which invited in more European missionaries, merchants, explorers and led to the eventual annexation and exploitation of the African continent by the Europeans that goes on till today.

Growing up in remote Africa and learning from history how slave traders and slave raiders penetrated the interior of Africa to get slaves made me wonder why the natives couldn't unite themselves to fight the invaders who came to capture their strong young men and women and plunder their future. It could have been easier perhaps to fight them off, though other circumstances would have permitted the slave hunters to be successful on their missions.

As children, sometimes, on our way home from the farms, we would run ahead of our parents who couldn't keep pace with us. Other times, we left the farm earlier than the adults so as to return home and fetch water from the stream or firewood from the forest. These roads from the farms to home were bordered by walls of thick bushes. At times I wondered what would happen if a wild animal suddenly emerged from the bushes to attack me. I will be helpless because there were no farms or houses nearby to hear my cry for help and rescue me. "Maybe, it is such circumstances that the slave hunters had benefited from and attacked people in the hinterlands in order to capture slaves for the Whiteman." I would imagine.

Though there were some forms of business and interaction with distant communities before the coming of the Europeans, Africa's activities included mainly farming and grazing in the hinterlands. Trade was somewhat limited due to transportation challenges. Trade by barter existed in some parts, practiced between closer villages or communities. The people along the coastal lines had rivers and seas as their sources of livelihood, fishing being their main occupation. These trends have continued with different people for centuries as Africans far from the coastal regions still depend heavily on grazing and farming, while the people along the coast still have fishing as their respective occupations.

History tells us that the Europeans used these coastal indigenes to launch raids into the hinterlands to capture slaves. "I could have been captured on my way from the farm or my friends and I must have been captured in the forests when we go to fetch firewood" I thought of it while battling with tall grasses brushing against my face as I returned from the farm with a heavy load on my head or when out in the woods. Coastal middlemen sponsored in-land lords to raid and bring them slaves for the Europeans goods, a situation that we could have fallen into if such raiders were present during our growing up days.

"Imagine slave hunters still moving around catching slaves, what will become of us if they ambush us here?" I once asked when we were far in the forest, far from home fetching firewood.

"I will kill one of them before they take me away, I will cut whoever comes to catch me" My cousin Muluh retorted with much confidence, displaying his well sharpened machete as we all burst into laughter.

Our grandparents told us war time stories between different tribes, mostly tribes close to ours. Our tribe had as well fought against other tribes. I remember my grandmother recounting how our people once fought a couple of neighboring tribes. "Given that the Bamombous and Balis lived downhill below us, we have always had an advantage over them since we could attack them from above." My grandma told us. "The Bali people, after previous defeats avoided fighting us because of our strategic position" grandma would say with some pride "after our people had defeated Bali in a war." she continued. To this day, there is a folk riddle which reminds us of this victory.

Land was an important asset back then as it is today for farming and grazing. Most wars between kingdoms came about as a result of perceived encroachment into land by a neighboring kingdom or state. In recent times, this cause of conflict hasn't gone away. Among our people, there are family lands but there are also crow-lands on which community farming and grazing are practiced.

A group of people from a neighboring village to ours, Allah-Chu, were rounded up by my village people, Buchi and taken to our chief's palace. The chief, Mokom Ndi IV on his way from the city, was informed of the happenings. He immediately requested the civil administrator, the Divisional Officer (D.O) to accompany him to the village. Arriving at his palace accompanied by the D.O and a few security officers, some men and women from Allah-Chu were held captive in the courtyard of the Buchi palace, my own village palace.

My village people had complained to the chief about the neighboring villagers encroaching on farmland beyond their boundary. They had warned the neighbors about farming on the said land over and over. On the day they were whisked to our palace, they were accused of continuing to farm on the land which wasn't on their side of the divide. The case was resolved by our chief and the D.O, the captives from the village across were set free. It was a situation that could have provoked a full-blown war if the officers, civil administrators had not been present. Back in the days with no central government, the other tribe upon getting news of their people captured, would have immediately retaliated. So many of such inter-tribal conflicts have been solved by the civil administrators over the years as we grew up.

"Given that some tribes were weak, always defeated at war, they willingly accepted the Whiteman when they came.The Europeans had brought weapons which were more powerful than the indigenous people's spears and arrows." Our teacher explained to us. "In addition to other goods, guns were introduced which they used in fighting the natives as they started moving into the hinterlands. Zeinab Badawi in the BBC's history of Africa explained that "The Benin warriors' machetes, spears and arrows were no match for the British sophisticated rifles and cannons" as they were determined to take over the Benin kingdom.[2] As tribes were attacked and their resistances failed, some of the natives fled further into the hinterlands. As they moved inwards, they passed the news about the Whiteman and his superior weapons. Some weaker tribes rather welcomed the Whiteman who protected them against more powerful tribes." Our teacher told us.

The Bali King, Galega I, whose kingdom shares borders with my clan, Pinyin, happily welcomed the German explorer Eugen Zintgraff and on August 23th of 1891 signed a peace treaty with the German explorer. In this way, his kingdom enjoyed some privileges like protection over other tribes.

As far back as the early 1500s, the Benin empire (today's Edo state in Nigeria) under the rule of Oba (King) Osigie, allied with the Portuguese military to conquer the Idah people and other kingdoms. His alliance with the Portuguese helped him consolidate and even expand his kingdom before it fell to the British during the famous Benin punitive expedition of 1897 under the rule of Oba Ovonramwen. The empire fell to the British as Oba Ovonramwen was deposed and exiled to Calabar.[3]

The European explorers easily discovered the Africans' weakness immediately when they arrived at the coast. It was partly this weakness of some Kings and their subjects which made them see the Europeans as a means of them gaining leverage over their fellow Africans. This same weakness has gone on for as long as the Europeans have been in Africa, hundreds of years and counting, they still use fellow Africans to subjugate other Africans. Africa has ever since been a prey to the Western world to a greater extent, and other parts of the world to a lesser extent feed on.

Through historical records, I have come to learn that the Portuguese who were led by Fernando Pô and who first docked on the coast of

Cameroon were the same sailors who sailed through the West African coast and reported to the Europeans. Fernando Po explored the vast Gulf of Guinea, its potentials and then returned to Europe with the good news of a rich land full of riches to his people.

Journalist and explorer Henry Morton Stanley sailed to the west African coast around the same time as Fernando Po in 1482. His discovery of the Congo caused him to report to the British who were reluctant to get involved, then he went to the Belgian king, Leopold II who later on turned the Congo into his private colony. King Leopold committed some of the most heinous crimes in history against the people of the Congo. Some scholars opine that it was a genocide as the estimate stands between ten and fifteen million Africans killed and a great number left with amputated limps[4]

The easy entry and annexation of one European country into Africa was a certain invitation to their neighbors to follow suit. It is a fundamental human tendency to follow the actions of others, especially when such actions result in great success. Each era always has something that one person or nation initiates and later on others join in competition. In the era of the scramble for Africa, national pride amongst the European countries was at stake.

"Fernando Po's voyage to the coast of West Africa in 1472 marked the beginning of the acquisition of African land" my secondary school history teacher told us in an African history class. He continued, "The frontrunners in African colonization were the British and the French. On the set, the Germans were not so interested in the acquisition of colonies in Africa. However, later on,

Bismarck, Germany's chancellor at that time, gave his famous speech on Germany's "place in the sun" in which his vision for Germany as a colonial power became the ideological foundation of German naval expansion which began shortly thereafter. They became the 3rd European country with multiple colonies in Africa amongst some states that had just one or two colonies. The Congo free state was the personal property of the king of Belgium, King Leopold II." our teacher said with some sarcasm.

Though not so united at the time, the colonialists displayed a high degree of intelligence which I live to applaud them for. Just like the big cats in the wild, they always advance in some form of unism though they all never end up eating the same amount of meal. Following what I have learnt about prides of lions, a lioness is more accurate in hunting than the male (lion). But in most cases, after a pride catches a prey, the males feed first before the females. Though they all sacrificed towards the catch, the lioness being the most skillful in the hunt, they do not end up having the same share. Still, they continue life in that manner because it is more productive.

The desire for colonies on the ''Dark'' Continent as most Europeans viewed the African Continent, coupled with their national pride pushed the Europeans to scramble for colonies in Africa. However, going forward they had to be tactful, not jeopardizing the general mission, capturing Africa for themselves.

In 1884, the German chancellor, Otto Von Bismarck organized a conference in Berlin Germany on how the European powers had to go about colonizing Africa. The Berlin West African conference of

1884 is one subject in history that we received with lots of resentment. Just like the pride of lions who never inform a herd of buffaloes or wildebeasts about their mission, the Europeans met at Berlin and planned their ambush on their targeted prey, Africa, without the knowledge of the Africans. The fate of the entire continent to be partitioned, Africa, was initiated and as a fallen prey, they all devoured the vast and rich continent. They have continued reaping/ripping off the African continent ever since, never relenting in devouring their coveted trophy.

At times, in our history classes, everybody including our teachers couldn't help but become very angry. Each time I hear such words as the Berlin West African conference and slavery or slave trade, the same raw emotions of anger come rushing back. "Yes, they went and sat in Germany and agreed on how they will effectively conquer us" the teacher emphasized as to make us really feel hurt about colonization.

In 1884, the Western European powers portrayed an act of intelligence, one compared to that of the big cats in the wild. The explorers saw this giant prey, reported to their people who in turn put their intelligence at work. Though they all didn't end up with the same proportion of the prey, they went in as one unit and the prey continues to be expended till date.

In 1884, the Western European powers portrayed an act of intelligence, one compared to that of the big cats in the wild. The explorers saw this giant prey, reported to their people who in turn put their intelligence at work. Though they all didn't end up with the same proportion of the prey, they went in as one unit, captured the prey and it continues to be expended till date.

CHAPTER 2

MANIFESTO FOR COLONIALISTS

CHAPTER 2

MANIFESTO FOR COLONIALISTS

The intelligence of the Europeans went much further and deeper than the meeting in Berlin of 1884. They employed the highest tact and ingenuity to ensure their catch had no way of escape. European rulers such as the Belgian King, Leopold II instructed Belgian missionaries in particular and the European missionary body sent to Africa by colonialists how to relate with people in the colonies.

In 1883 King Leopold II of Belgium wrote a letter to Belgian Christian missionaries being sent to Congo. These Christian missionaries would eventually become the spearhead of Belgian colonialism only to be followed by Belgian traders then the Belgian army. His letter read thus:

"Reverends, Fathers and Dear Compatriots:

The task that is given to fulfill is very delicate and requires much tact. You will go certainly to evangelize, but your evangelization must inspire above all Belgium interests. Your principal objective in our mission in the Congo is never to teach the niggers to know God, this they know already. They speak and submit to a Mungu, one Nzambi, one Nzakomba, and what else I don't know.

They know that to kill, to sleep with someone else's wife, to lie and to insult is bad. Have courage to admit it; you are not going to teach them what they know already. Your essential role is to facilitate the task of administrators and industrials, which means you will go to interpret the gospel in the way it will be the best to protect your interests in that part of the world. For these things, you have to keep watch on dis-interesting our savages from the richness that is plenty in their underground. To avoid that, they get interested in it, and make you murderous competition and dream one day to overthrow you.

Your knowledge of the gospel will allow you to find texts ordering, and encouraging your followers to love poverty, like "Happier are the poor because they will inherit the heaven" and, "It's very difficult for the rich to enter the kingdom of God." You have to detach from them and make them disrespect everything which gives courage to affront us. I make reference to their Mystic System and their war fetish – warfare protection – which they pretend not to want to abandon, and you must do everything in your power to make it disappear.

Your action will be directed essentially to the younger ones, for they won't revolt when the recommendation of the priest is contradictory to their parent's teachings. The children have to learn to obey what the missionary recommends, who is the father of their soul. You must singularly insist on their total submission and obedience, avoid developing the spirit in the schools, teach students to read and not to reason.

There, dear patriots, are some of the principles that you must apply. You will find many other books, which will be given to you at the end of this conference. Evangelize the niggers so that they stay forever in submission to the White colonialists, so they never revolt against the restraints they are undergoing. Recite every day – "Happy are those who are weeping because the kingdom of God is for them."

Excerpts from King Leopold's letter according to the African Exponent.[1]

King Leopold's letter to the missionaries was discovered in a Bible in 1935. Moukouana Muikwani Bukoko, a native of Congo is quoted to have bought a secondhand bible from a Belgian priest who had forgotten the letter in it which dated as far back as 1883 from King Leopold II of Belgium.

Leopold II's letter was the foundation on which the Europeans would mainly build their policy towards the Continent; the 'inferior' African to the 'superior' European; the 'dependent African to the 'benevolent' Whiteman. Using the Bible as a weapon against the

Africans was a very effective strategy that continues to subdue the African mind till the present day.

Once the predator has conditioned the mind of the prey that they are inferior; that they are never to aspire to be better than what they are; so that they cannot challenge why they are constantly preyed upon; that they are weak, though they have strength; that they should humbly accept all that their masters hand to them; that their status quo is destined by creation or God, then the possibility for the prey to ever rise up to a better life is shrunk to almost zero, it becomes an illusion. The prey can only challenge their status quo if they realize that their present or even past reality is not what it is supposed to be.

For hundreds of years, the land of Africa has been under the control and exploitation of the European powers that colonized it. Much has been said and written about the enormous riches that lie beneath African soils and above in its flora and fauna. Such stories about the riches of Africa have oftentimes made my heart swell with pride, eyes filled with tears of joy for belonging to such a wonderful and blessed land. However, the cruel reality on the ground quickly overturns the joy. The squalor that characterizes a majority of Africans cannot be reconciled with the riches and potential of the land, from the Cape to Cairo, from Darker to Mogadishu.

In 2011, I made a journey into one of the richest places in Cameroon; the East Region of Cameroon which is part of the Equatorial Forest. Never in my life had I seen such rich forests before, with trees so big that it takes a trailer truck to carry a single tree trunk to the port for exportation. Driving from the capital city,

Yaounde, I enjoyed the tarred road for a while, until three hours later, the luxury of a tarred road came to an end. We arrived at a small town called Abong-Mbang. Dust immediately welcomed us as we exited the tarred road. It was a former district headquarters when Cameroon was a German colony. Remnants of German colonial presence were still very visible in this part of Africa. "When the Germans were here, they built this prison". A local told me who became my guide for the few hours that I spent at Abong-Mbang. He took me to the administrative and residential quarters, "These are all memories of what the Germans left as colonial masters, no renovation on any of the infrastructure" he continued with some disappointment on his face.

In a whole lot of places that the Germans were present in Cameroon before losing the first world war in 1919, their infrastructure still stands till date, even the railroads that they built still serve the locomotive engines that ply the rails in Cameroon.

After a night of rest in Abong-Mbang, our journey into the Baka world started. A few miles from the little town of Abong-mbang, the Equatorial Forest welcomed us with its grandeur of giant trees. The giant trees guarded us on both sides of the dusty road, their huge branches and leaves towering above us as if they were super-sized guards watching out for whoever entered the land. Dust particles engulfed the air as the lead car before us provoked the dry tiny earth particles, forming a brown cloud that blurred visibility. Pastor Bangbe Claude was our lead driver in a small Mitsubishi Lancer. He was no stranger to the terrain as this part of Cameroon is where he comes from. Red lights flashing from tail lights (brake lights) on

Pastor Bangbe's car ahead penetrated the brown cloud as my guide along the narrow road with lots of bends and twists which required slamming on the breaks every now and then.

After driving for about forty miles, we arrived at a locality called Mindourou. There, lots of trucks stationed with their trailers tied on the backs facing the same direction we were headed in, while others with the trailers connected, with huge logs of timber placed on the trailers designed specifically for transporting the giant trees headed in the opposite direction towards us, to the city. This junction with lots of activities was the business hub of Mindourou; a resting place for drivers as well. I had not seen a single truck ever since we left Abong-mbang, but here, they were all lined up, some with logs facing the direction we came from and others without the giant tree trunks facing the direction we were headed. We stopped to relax and to wait on our racing colleague who couldn't keep up with us with his old truck. The dusty place was a hive of economic activity. We spotted a 'soya' joint. Soya is meat roasted with spice and then sliced into small pieces, a common treat in West Africa. We walked over to the little makeshift hut and were welcomed warmly by the soya man. It turned out he was a man from my region of origin. My poor French language caused him to ask where I hailed from. He was even happier to see someone from his region as it was not every day that he was privileged to talk to someone from his part of the country.

My discussion with the soya man went on for a few minutes during which I learned that the junction never slept as truck drivers going both ways used it as a rest area.

"Every day, millions of Francs cfa leave this region, but the people have nothing at all in compensation for their trees that leave the land" my new acquaintance told me. "These are millions lined up on these trucks my brother, what you see is nothing as they have not started returning. This goes on every day of the year, even on public holidays" the soya man said with the same frustration that many Africans always show at the plundering of their natural resources.

His words stuck with me as I embarked on the last lap of our trip, the road dustier than before and view becoming almost impossible. "Use the windshield fluid to clean the windshield" a voice in my head hinted to me. "No, the dust is way too thick, it will cause mud if you spray the fluid, it will only make a bad situation worse" Another voice countered the first. "Just turn on the wipers without spraying the fluid" yet another voice said to me. There was total silence in the car; sure no one wanted to distract me given the type of terrain we were driving on. I turned the wipers on, the first wipe made the windshield so clear, I was surprised to see how clear it was, I'd been driving for that long not knowing I could do it. In less than no time, the dust had covered it again. The dust was falling on the windshield as if it was raining dust from the skies, but it was no problem this time around because my wipers cleaned the fine dry dust with no issues.

We drove for over an hour, going through small villages as children rushed to the road at the sound of a car, waving at us and shouting words we couldn't decipher. Our windows all rolled up to prevent the dust. They reminded me of when we were young, we would rush up to the roadside when a car sounded from a distance, waving, and shouting "buy me biscuits, buy me chocolate" just any nice thing that would come to mind at that moment though our requests never yielded any results. I wished I could stop and hand these children something, but I couldn't. Driving in this part of the country was risky, not only to the driver but to the locals as well. Their settlements were so close to the road with children playing in dusty yards in front of their thatched houses. In some places, I drove past before noticing someone was standing beside the road. Despite the truckloads of wood exploited from here, the powers in charge could not afford to enlarge the road for the exploitation activities.

"If only you know the types of machines that are in these forests, you will feel sorry for the people in this place. Their only mission being to bring out the trees for transportation to the city, zero interest to improve the lives of the people they cut their forests" Mama Mariam, pastor Bangbe's wife broke the silence in the car. This was her husband's village. "We will soon arrive at Pastor's mother's house" she continued.

We made a sudden stop, I found out one of the cars ahead of us needed some attention. At the rest area, the third driver had refused to stop when he caught up with us since we were faster than him. Deep in the forest, his old truck broke down. The night was still early, about 7p.m, beside the sounds coming from us, sounds also came from the forest; some strange and some familiar, some from afar and some just close by. The Congo basin is home to diverse animals, some of which I have never seen because they are not present where I came from. I had only visited the forest through the eyes of TV channels like the BBC and National Geographic that do documentaries about life in the Equatorial Forest.

Off we went, leaving behind Constant the truck driver and his passengers. Thick darkness had taken over from the daylight. The forest canopy seemed to have made the night even darker than normal. Though the headlamps of our Toyota Avensis were outstanding, they could barely penetrate the darkness under the giant trees, it was frightening. Later on, I realized it was the darkness coupled with dust that made it even darker and difficult for the car's lights to penetrate. There were oncoming flashlights from the opposite direction of the road. Pastor Bangbe slowed down, creating more room for the approaching vehicle by shifting into the bush on the right and turned off his headlamps. I did exactly as he did. "We have to obey the big boys around here" I said, we all in the car laughed at my comment. It was a wise decision to respect these heavy-duty trucks. We could feel the warmth in the car coming from the trucks' beams as they drove past us. We'd respected every single one of them until we arrived at Lomie, our final destination for the long journey.

Somewhere in the eastern part of Cameroon, in the Equatorial Forest we were out of touch with telephone network signals. Our short stay in Lomie was great, experiencing different food varieties and coming into contact with the pygmies of the Congo basin with their great hospitality.

The people in the forest depend greatly on hunting animals for consumption. I came in contact with a few Bakas and ate some meat which I did not know which animal exactly it came from. Among the Bakas, I suddenly seemed to appear like a very tall fellow though not the case when I am amongst other people. "Here, the people don't depend on any outside medication, maybe because there is none. Every sickness has a cure, and it is from the forest" a man told me. I liked their way of life, they seemed to have no worries whatsoever. "Even the white man comes here and learns medicine from these people, they learn it for free and take these African medicines to their countries and make money from it," the man whose name I never got told me. They lived a simple but fulfilling life. The standards of the world seemed not to have any effect on them. As long as their forest stays in place, they will have crops for food tomorrow, they will have animals which they hunt for food for tomorrow, they will have their clothing for tomorrow, and they will have a place to rest their bodies for the night. All these may not be up to the standards of the world, but it is their way of living, one which they've preserved against all odds, resisted to give it up in place of the world's standards. My greatest fear was what would become of them if one day their forest is wiped out given the extent to which exploitation is being carried out. Their way of life and possibly their existence is endangered.

The Congo rainforest in Central Africa is the biggest rainforest after the Amazon rainforest in South America. According to the World Wide Fund for Nature, known commonly as WWF, the Congo Basin spans across six countries: Cameroon, Central African Republic, Democratic Republic of the Congo, Republic of the Congo and Equatorial Guinea and Gabon. There are approximately 10,000 species of tropical plants in the Congo Basin and 30 percent are unique to the region. [2]

At an early age, I had learned of the importance of these rainforests to the survival of the human race and planet Earth. As we left Baka land early Monday morning back to the city, we faced the same challenges we had faced on our way there, I couldn't stop my eyes turning to look into small communities we passed along the way. Nostalgia set in as I thought of my experience as a kid growing up in my village, Buchi in the Pinyin clan. Early in the mornings, we would go out to tether goats in the fields, to fetch water or on other errands.

Many years have gone by, and I still think of my trip to Lomie, the Pygmy land with mixed feelings. I remember driving through the difficult terrain. Other times, I have thought of the never ever-ending exploitation of the rainforest by greedy insatiable companies from far away, without any regard to the people whose livelihoods depend entirely on these giant trees. Such journeys always fascinate me and have birthed in me a dream of traveling the length and breadth of Mother Africa, to get first-hand experience of all what goes on in her bosom as well as the challenges her children go through every day.

A cursory glance at the Continent of Africa shows that the collaboration exhibited at and after independence from the 1950s till present day by the European powers and the Western world reveals that King Leopold's letter couldn't have been taken lightly, and that the fallout of the conference held in 1884 in Berlin by the colonial powers is still much practical on the Continent today.

The West has been the predator on the attack, laying all sorts of ambushes against the African prey who for centuries lack the wit to stand as one and fight back.

Once the predator has conditioned the mind of the prey that they are inferior; that they are never to aspire to be better than what they are; so that they cannot challenge why they are constantly preyed upon; that they are weak, though they have strength; that they should humbly accept all that their masters hand to them; that their status quo is destined by creation or God, then the possibility for the prey to ever rise up to a better life is shrunk to almost zero, it becomes an illusion. The prey can only challenge their status quo if they realize that their present or even past reality is not what it is supposed to be.

CHAPTER 3

THE GREAT DECEPTION

(Decolonisation & Independence)

CHAPTER 3

THE GREAT DECEPTION

(Decolonisation & Independence)

Nothing lasts forever, they say. To everything, there is time for a start as well as a time for an end. History tells us of great empires that ruled the world in different eras. When I learnt of Alexander the Great in the movie Alexander which depicts the rise of Alexander, the young boy who became king of one of the Greek kingdoms of Macedonia, I was in awe of the young boy who could build an empire at such an early age, and for centuries his name and fame never waned. "What actually motivated this kid king with such ambitions to conquer territory after territory?'' "Did he for once think all his conquests will someday cease to be part of his empire?" These are questions I ask myself when I read or watch videos about Alexander the great.

My secondary school history teacher, Mr. Atud, taught us about the bravery, intelligence and even cruelty of the great king Shaka of the

Zulus in the Southern African region. In this class, I sat on the last bench, I envisioned myself as I listened to stories about Shaka, his army unrivaled by any in that part of the world at the time. Probably he had envisaged capturing and subjugating all the kingdoms around him, and this has been the same with all empires and conquerors through the ages.

After hundreds of years of being under colonial rule and domination, the African people started demanding for self-rule in the mid-20th century. Violence erupted in most of the territories occupied by the Europeans. No occupier ever willingly cedes control of territory they are comfortably exploiting. The occupier's wish is to keep the oppressed people under control for as long as they can, by providing unending reasons for oppression. All oppressed people, as history puts it, have come out of oppression by rising up and challenging the oppressors. Africa's case was no different as in the second half of the 20th century; the spirit of nationalism and Pan-Africanism fueled a few individuals who spurred the people's spirit of determination to challenge European hegemony and exploitation.

My pride as a descendant of these African heroes surges when I read of how they rose across the board and challenged or even fought colonial rule. If there is one thing that a pride of lions may fear in the wild, it will be the day a herd of buffalo's team up as one and go after the lions, undoubtedly, their vast number matched with their brute power will crush the lions.

Certainly, with calls to end colonialism echoing from across Africa, the Europeans feared they had no chance to continue having their hold over these people. If the Africans chose, they could rally themselves and stand against the whole of the superior force of the colonizers, and if they did, not even the entire European armies could withstand, given their numbers.

Independence was not a choice by the colonizers as it has never been a choice by the oppressor to let free their prey, it was rather an unconditional option given the European colonialists by the colonized people of Africa.

"Although this independence of the Congo is being proclaimed today by agreement with Belgium, an amicable country with which we are on equal terms, no Congolese will ever forget that independence was won in struggle, a persevering and inspired struggle carried on from day to day, a struggle in which we were undaunted by privation or suffering and stinted neither strength nor blood". [1] These words are an excerpt from Patrice Lumumba's speech at independence, reminding his compatriots and the rest of the world what it took them to get their desired independence.

As a result of the great pressure put up by African states, Ghana emerged the first from Sub-Saharan Africa to gain her independence. Ghana's independence on the 6th of March 1957 paved the way for the rest of the African countries to gain theirs.

Independence for the colonies meant the European powers were going to lose a lot of the very resources which had influenced them to colonize a whole continent to which they had had free access for centuries; they were not ready to let go of all this.. Western governments had thrived on the human and material resources of

Africa for centuries, in no way were they willing to put these behind them and move forward.

While Africans sat back in the comfort of their various countries and celebrated their victories in obtaining freedom, the intelligence of the predators immediately went to work on schemes to keep the Africans under continuous rule, perhaps this time not as directly as before. They never imagined giving up on the colonies, even making it clear to the people that they were not ready to let go of them.

The great French general who led the French against Nazi occupation during the second World War, Charles de Gaulle went on to become French president during the early decolonization days. Most of the French colonies became independent during his presidency. The first was Algeria that had waged a bloody war of independence against France. He knew that granting independence to Algeria would anger some French people, particularly the settlers who had to return to France following Algeria's independence.

In 1958, following the choice in a referendum given by De Gaulle in which the colonies were to become more autonomous in the new French community or immediate independence, Guinea Konakry under the leadership of Sekou Toure was the only African colony not to choose more autonomy within De Gaulle's new French community. Guinea's decision to become completely independent angered the French, their colonial masters. The Washington Post observed "how brutal the French were in tearing down all that they thought were their contributions to Guinea."[2] -

The response given to Guinea by France was a stern warning to the other French colonies in Africa should they do what Guinea did. The French removed everything they possibly could, including light bulbs, and what they couldn't carry away such as medicines, they rather got them burned than to be left for Guineans.

In the height of the demand for independence, then French justice minister Francois Mitterrand in 1957 supposedly made a statement which has been debated by many if ever such words were uttered by the statesman or not.

"Without Africa, France will have no history in the 21st century". Supposedly Francois Mitterrand said in 1957 while he was still minister of Justice, proposing France's pathway with the soon to be newly independent African colonies. His proposal was said to have been reported in the printing press by the 'Politique Francaise'. Mitterrand ended by saying "already, France knows how much Africa is necessary to her." Mitterand later went on to be the longest serving French president, from 1981 to 1995.

Coming from one of the countries that has suffered and continues to suffer from the French policies in Africa, these words coming from Francois Mitterrand make me cringe. "How could someone make such a revelation and continue to rip us off our riches, after having ripped us for hundreds of years?" I asked. More like a lion telling a group of brainless buffaloes that their existence cannot be possible without the lions killing them for food.

By 2008, I was a young man who would have graduated from university if all went well. I loved to follow international politics as I had grown old enough to comprehend happenings around the world. Getting into university was the age at which most young people in Africa who went through school dreamt of improving their lives and those of their families. Dreams that seem so near but at the end of university studies, they become so far. Dreams that often look so real at matriculation ceremonies but never come true after graduation ceremonies. We had come to know that leaders in the West had policies that directly affected our dreams. We wished for some current Western governments to be voted out of office so the incoming ones may bring in place certain policies that will affect us too or put pressure on some of our leaders to make way for new leaders to pilot the countries.

In 2008, it was alleged that Jacques Chirac, outgoing French president said: "without Africa, France will slide down into the ranks of a third world power." [3]

Although claimed by some as a conspiracy theory, some publications like the Atlanta Black Star and reddit which are all online publications confirmed that such words were actually uttered. Also, a French publication, deplomatie.gouv.fr is said to have done the publication. This statement animated endless debates among us youths and has remained one of the statements by which Africans debate about Franco - Africa relations.

"If we think France is willing to let go of Africa, we are barely wasting our time and daydreaming" one person said.

"Let me tell you guys something, so long as our dictators here in Africa pay allegiance to France and ensure they have their way in the resources France needs, the only thing that will take them from power is death. It doesn't matter which president is elected into office, their foreign policies towards Africa remain the same" An older man interjected. His words were demoralizing to me. Young men like myself often found comfort in the hope that the long-term leaders will leave power even if we knew such promises were false.

In the year 2004, happiness engulfed lots of people in the West African state of Cameroon and people celebrated on the streets. The news was that the long-time ruler of our country, Paul Biya had died in Europe, in a hospital in Switzerland. It was news that most Cameroonians had been waiting for, some for their entire lifetimes- the death of a dictator who served France more than he did his fellow countrymen. "Wuna don hear say Paul Biya don die? (Have you people heard of Paul Biya's dead?)" A voice in the dark announcing the death of the president with so much delight and passion. It was no secret that a majority of the people of Cameroon, after failing to remove the president through the ballot box, had wished him dead after over two decades in power. We jubilated at the news, and prayed to God for it to be true. My body felt as if some heavy weight had been taken off my shoulders. The following days came with lots of uncertainty, speculations about the president, prayers being offered for the news to be true. Unfortunately, at the end, the news was confirmed as a mere rumor as the president has continued to be president till date.

Of course, many years after the former French president's declaration, the French policy towards the continent of Africa has not changed one bit. Successful French presidents keep telling the people how important Africa is to France and how the relationship needs to get even stronger.

"France, along with Europe, would like to be even more involved in the destiny of Africa..." the new president of France, François Hollande in 2013 came up with his own vision for Africa in relation to France and Europe at large. As if they have ever been less involved. 4 -

In 2007, one of the world's brightest brains, the man credited with the discovery of DNA after his double-helix structure of DNA, Dr. James Watson, during an interview with Britain's Sunday Times stated that white people were more intelligent than black people. He further went on to say that he was "inherently gloomy about the prospect of Africa" because "all our social policies are based on the fact that their intelligence is the same as ours, whereas all the testing says, not really." Dr. James Watson went on to declare "I wished everyone were equal. People who have to deal with black employees find this not true." James Watson continued his testimony of blacks and Africans as a whole. 5- Dr. Watson's revelations caused an uproar, not only amongst Blacks but across all other races.

I tried to gauge young people's reactions in discussions when I first listened to the reporting about Dr Watson's assertion over the BBC in 2007. With my cousin Menyam Destin when I told her about what I heard on the news. She immediately stopped what she was doing, her face dropped.

"There is no way a Whiteman will be more intelligent than me" Menyam said with unmasked anger in her tone.

"Relax, it is not me who said the Whiteman was more intelligent; I am simply repeating what I heard on the news today". I tried to calm her as I realized how disturbing this was to her.

"Who did you say the name of the doctor was again?" she asked with a tinge of anger in her voice still..

"Dr James Watson, I heard on the BBC that he once won the nobel prize" I replied.

"Well he is just another racist, only racists will make such statements" my cousin concluded.

Just like my cousin, other scientists and researchers termed James Watson as racist, arguing that his revelations had no base. The scientific community distanced themselves from James Watson and isolated him.

"Watson's beliefs stemmed from some geneticists who believed that there were relationships between IQ scores and genetics" a geneticist disclosed on CNN after the remarks that James Watson made.[6]

"There is no scientific basis suggesting that the two are causally linked, or that IQ scores are even a reliable measure of intelligence. The most obvious explanation for differences in IQ tests is environmental factors affecting a person's upbringing" Joseph Graves on Anderson Cooper in 2007 explained on CNN.

Whatever the statements from the scientists be, a people cannot thrive if they let others tell their stories.

"People will buy you the same way you sell yourself" Nathalie Yamb, one of Africa's critical minds that speaks up against French evil policies in Africa once said. She shares on her twitter handle regarding how African leaders present Africa on the global stage. She first caught my attention when I heard her talk in Russia, at the Russian - African summit in 2019. She strongly condemned the fact that France holds the destiny of millions of people in her hands.

If Dr. James Watson's statements were wrong, the African people and their descendants the world over have failed to prove it. We Africans seem to rather confirm as we continue to lag behind in all spheres of life. According to Nathalie Yamb's words, the price that Africa receives is equal to the price tag that she's placed on her. Until we realize our full value and tag ourselves with a higher price, we will continue to be regarded as such.

A predator in the wild needs to survive, it survives on the prey that doesn't have the intelligence of the predator. A cheetah in the wild goes after a group of wildebeests and succeeds to bring down a wildebeest twice its strength and size. The TV footage made me applaud the smaller animal for its bravery and courage when I followed this. While the rest of the wildebeests ran about still in fear of the lone cheetah, it was already pulling its dead prey up a tree where other bigger cats or scavengers couldn't reach him. The film maker remarked that the cheetah is one of the most skillful big cats in the wild when it comes to hunting. It is fast and swift in its attacking skills; it can chase prey for much longer distances than rivals like the lion and tiger that rather rely heavily on teamwork. Still, I feel sorry for the preys that lack the intelligence to recognize the power in their numbers. Though the cheetah is alone, smaller in size than the prey is, it still succeeds in using them as food.

The bigger animals in the wild, just like Africa, have presented themselves as foolish and vulnerable. As a result, the predators simply take them for what they have presented themselves to be.

In our groups of young debaters, we often talked of how France is the weakest of the European powers. Sometimes, during our random discussions we bragged on how Africans saved France from an embarrassing defeat during both World wars. Yes, our people helped the French, especially during the Second World War and their efforts couldn't be taken for granted in defeating the enemies. France, though, sees herself like the superior guy in the wild who feeds on the same people who helped her. Maybe these Africans who fought for France never did so out of their own volition, if France has succeeded in exploiting them even after independence, then their wartime services to the Republic of France were just in obedience to what the master required of them, in relation to how the master had conditioned their minds. Even the French have recognized and confessed that they cannot do without Africa, just as the cheetah cannot survive without wildebeests and deers in the wild.

I can't stop wondering what goes on in the minds of the big cats in the wild. "Surely they constantly reassure themselves that the bigger and stronger guys are not as intelligent as they are, though muscular and more powerful, what matters in the wild is intelligence" They make sure that they pass these principles to their young. A lion trains its cubs on a baby deer. Their parents capture the fawn alive and hand it to the cubs. They then sit back and watch as their babies practice on the baby deer. Same thing with tigers and cheetahs, young prey is captured by older members of the clan, handed to the young ones who are given the opportunity to train on. Maybe the other fellows never believed in themselves as having any possible means to challenge those that feed on them, their children grow up in the same ignorance and therefore remain as food to the other type of people.

The relationship between Africa and France, Europe and even the rest of the world can be said to fall under the words of James Watson "their intelligence is not the same as ours", the reason why "Without Africa, France will slide down into the rank of a third [world] power." and they keep exploiting Africa to stay afloat like the rest of the so-called Great powers, yet Africans stay in silence and die in misery. "What else will make 'a people' come to the realisation that they deserve better?" Maybe we all know we aren't any better than what and who we have always been. Maybe we tell ourselves our current state is who we are meant to be because it is who we are and as such will remain like that- the most undeserving people of the world.

CHAPTER 4

THE STORY UNTOLD

(How they did it)

CHAPTER 4

THE STORY UNTOLD

(How they did it)

One of the greatest deceptions of European colonialists was to make Africans believe that they had gotten the much cherished, hard-fought for independence, but never fully delivered it. The master never ceded his powers, he only changed tactics. Africans were going to have leaders such as presidents and or prime ministers, but it was made clear to everyone that the masters from Europe would give orders and the servants would follow. They never granted independence to the Africans because they felt it was time for them to let the Africans take control of their own destiny; it was a change of terms ostensibly to keep control over the people of the 'Dark continent.' Just as with the case of the slave trade, it was the industrial revolution which encouraged the Europeans to a greater extend to put an end to the slave trade. Their revolution in industrialization needed resources which came from Africa, keeping the African people in Africa to produce these resources was a better option than taking them to work on plantations in the Americas. However, they couldn't keep the Africans home without keeping them under their control, colonizing them was therefore the solution.

When colonialism was challenged as African nationalism increased, the West, unwilling to let go of the advantages they had enjoyed for ages from the 'Dark Continent' quickly hatched a plan to hastily grant independence to the various indigenous people.

The British had conceived a plan long before, to come up with a body overseeing her former colonies when they would eventually challenge British rule, thus the Commonwealth of Nations ensued. The French did the same with 'la Francophonie' to keep control over their colonies. Portugal on her part wasn't ready to let her African prey go as they too dreamed of a "Lusophone commonwealth" run from Lisbon.

"You don't ask a slave if he wants to be free, particularly when he is already in revolt, and much less if you happen to be a slave-owner."[11] Samora Machel, the face of the Mozambique revolution against Portuguese rule, rejected the Portuguese plan of a Lusophone commonwealth with this pitiful remark. Portugal didn't succeed as much as the British and the French in keeping her colonies under her control but attaining independence after a bloody war waged against the occupier (Portugal) was soon followed by a long civil war that lasted from 1977 to 1992. Mozambique later joined the

'Commonwealth of Nations' in 1995 which at the time was the only member nation that had never been part of the British empire.[2]

"For this independence of the Congo, although being proclaimed today by agreement with Belgium, an amicable country, with which we are on equal terms, no Congolese worthy of the name will ever be able to forget that it was by fighting that it has been won, a day-to-day fight, an ardent and idealistic fight, a fight in which we were spared neither privation nor suffering, and for which we gave our strength and our blood. We are proud of this struggle, of tears, of fire, and of blood, to the depths of our being, for it was a noble and just struggle, and indispensable to put an end to the humiliating slavery which was imposed upon us by force."[3]

he words of Patrice Lumumba in the speech at independence on 30th June 1960 to remind the predators present at independence that they had not withdrawn from them out of their good intentions. It was rather the sacrifices put forth by brave sons and daughters of the Congo that made it possible for the oppressed to have their freedom. Unfortunately, like most of Africa, this freedom they bled and died for was short lived because of the craftiness of the oppressor in complicity with our very own people.

In the years leading up to independence for many African countries, wars of resistance were strongly put up across Africa. It was a time when Africa exercised some level of cooperation amongst its people. It was the pressure put up by the African people that made the masters accept independence. Of course, the African people overwhelmingly surpassed the colonial masters in terms of numbers, holding up against the African people was a little foolish on the part of the colonialists. The natives realised that the 'Whiteman' was in no way superior to them, a notion fueled by their experiences fighting alongside white soldiers in the Second World War. With their empires across Africa, they did not have the manpower to suppress the Africans any longer. From The Gold Coast to Algeria, from Tanganyika to Mozambique; from Upper Volta to The Congo Free State, the spirit of Pan-Africanism filled the people. Their masters gave in to their demands with heavy hearts after bloody confrontations with the Africans.

Back to my documentary on TV, I jumped from the couch in celebration, as if my beloved soccer team, Manchester United, had scored a decisive last minute goal from one of David Beckham's signature free kicks. Finally, what I hoped could happen actually was playing before my very own eyes. Although lions hunt as a pride, they also succeed in hunting individually. A lion brought down a full-grown buffalo. As usual, the rest of the herd stood nearby watching. The buffalo cried in anguish as the lion choked the air of it, sinking its talons and teeth deeper every time the buffalo moaned. The herd mustered some courage to challenge the lone lion but retreated immediately at its roar. Then, they returned, cautiously approaching and were a few feet away from their friend and its captor. Before the lion realised, another buffalo had swung its powerful head, its curved dark horn caught the lion under the hind legs, and it went flying in the air.

"Yes, yes, yes, that's what I'm talking about" I jumped from the couch and swung my arms in the air in celebration.

"What's going on?" My sister asked from the kitchen.

"I'm watching the king of the jungle being defeated by buffaloes" I replied as my sister laughed at my reaction.

"This lion may not survive this impact inflicted by this single undercut" the narrator said adding to my joy. The film narrator sounded like a commentator commenting on a Mike Tyson match in which he had delivered his famous deadly undercut punch, knocking out his opponent to the canvas. I watched the lion limping into the bushes, showing signs of being hurt.

"If only these buffaloes can always be this smart" I murmured to myself, watching the buffalo that just got saved dangling from the pain it had endured from the lion's fangs.

"I'm not sure this lion is going to survive this, it's really hurt. A single strike from a buffalo kills many of these predators though the strikes are rare" the commentator explained the strength of the buffalo. "It is one of the strongest animals in the wild, you don't want to stand in its way when it's angry," he added.

Today, when I read about the great men and women of Africa who stood up to challenge the oppressors, I can't help but put them on a high pedestal. I have shed a tear sometimes when I learned these men and women didn't obtain as much education as some have today, yet they had the courage to challenge the system at the time. Any single country in Africa today probably has more educated people than the entire African continent had at independence. The International Institute for Applied Systems Analysis (IIASA) provides data showing how education has increased on the continent over the decades. [4]

Countless professors parade the corridors of African as well as western buildings today, but their education hasn't been able to challenge the problems posed by Western imperialists.

"We got educated, but we also got uneducated" I dare say about today's educated African elites.

The mindset that these revolutionaries and the people possessed was completely different from today's leaders and the greater portion of the African people. With little education, they understood the intentions of the colonial masters. Intentions to keep the African people under their thumb, intentions to have a continuous sway over Africa, intentions to continue the exploitation of Africa's resources.

"The real humiliating truth is that Africa is not free; and therefore, it is Africa which should take the necessary collective measures to free Africa". [5]

When I read through these words from Mwalimu, Julius Nyerere which he delivered in May of 1963 at the founding of the Organisation of African Union (OAU), I paused and wished I had lived in their era. "These were people with a lot of foresight," I said to myself. It seems we had just a few of them with this great intelligence and pride and longing for a free Africa.

It has not been long that I have taken much pride in researching and reading about African freedom fighters. A shallow history of a few leaders has been all that many of us were exposed to. Also, many artists from across the continent have mentioned some of these leaders' names in their songs.

When I first read about Samora Machel of Mozambique, his tough stance against the Portuguese and his contributions to the freedom of his country, the contributions of Tanzania towards the freedom of Mozambique encouraged me to make my next stop of reading there. Julius Nyerere took the relay from Machel and so the list continued to the Congo and to Burkina Faso to Ghana and on and on throughout the continent. Reading the sacrifices put up by the great men and women to tear down colonialism, I could only salute their efforts for taking the mantle to free a continent under occupation. I shed tears at some of the things they endured, and also felt more proud of them after knowing much of their efforts; then I cursed the successive leaders and our generation for allowing ourselves and Africa to fall back in the hands of the same masters that these noble men and women fought and died to liberate.

Sometimes I imagined myself standing with some of those who fought for Africa, I feel ashamed we have not continued what they started, ashamed that we chickened out and gave in to the occupiers.

"You people are still foolish enough not to realise that you all need to come together?" Julius Nyerere asks in my imagination as I meet them where they all gathered in agony over Africa's lack of intelligence.

"They have advanced greatly in education, some people having multiple doctorate degrees, but their intelligence has shrunk, look at them bearing all sorts of academic titles yet cannot protect the continent from the predators" Nkrumah speaking, sounding so dejected. Then I turned around to see the likes of Sankara, Lumumba, Machel, Traore just staring at me, depression and disappointment written all over their faces. The questions and looks from them all pierce my heart; I leave their gathering with a wound deep down in my soul as to how much we have led down a people who gave in their all as to free a continent from predators.

Africa unite, yeah!
Unite for the benefit of your people!
Unite for it's later than you think!
Unite for the benefit of my children!
Unite for it's later than you think!
Africa awaits its creators!
Africa you're my forefather's cornerstone!
Unite for the Africans abroad!
Unite for the Africans![6]

Bob Marley's words in the chorus of his song titled 'Africa unite' came to mind once as I lay in the field staring at the clear blue sky, the tall grass swaying back and forth at the command of the gentle blowing breeze. As I looked deep into the skies above, not only did I see the heroes of Africa who stood up against colonialism and imperialism, I also envisioned the host of African descendants assembled in sorrow regarding the state of their people all over the world. In the softest of voices that one can get, Bob Marley pleaded in his famous Africa unite hit. As I kept staring at them all concurring to the words that Bob Marley uttered, I noticed tears running down their eyes, weeping for the African race on the continent and the world at large.

"Could this be their tears dripping on me?" I asked myself after drops from the sky woke me up from my reverie. The blue skies had suddenly turned dark with drops of water from the skies. I couldn't tell how long I had been laying there as the sky had changed without me noticing. As I made my way home, Bob Marley's voice kept echoing in my ears, sounding even louder. "Unite for it's later than you think," "Africa you are my father's cornerstone"

The European powers, particularly the British, had undoubtedly foreseen that one day their great empire would disintegrate. S0 they envisioned another strategy to keep the former colonies tethered to them. The colonies of Canada gained dominion over their territories from the British and the French as far back as 1867.

Queen Elizabeth II, in her address to Canada on Dominion Day in 1959 during her tour of the United States and Canada in July of 1959, pointed out that the Confederation of Canada on July 1st, 1867 had been the birth of the "first independent country within the British Empire". She declared; "Dominion Day commemorates the birth of Canada as a nation and the first independent country within the British Empire. So, it also marks the beginning of that free association of independent states which is now known as the Commonwealth of Nations."[7]

As long ago as 1884, Lord Roseberry, the then United Kingdom prime minister, while visiting Australia, had described the changing British Empire as some of its colonies became more independent, as a "Commonwealth of Nations". If the colonies were to become independent, another means of keeping them under British influence politically as well as economically was hatched- The Commonwealth. [8]

This tactic used by the British inspired the French to come up with a similar organisation, (La Francophonie.) on March 20th, 1970, to keep their colonies under their control when they sought autonomy or independence. At the independence of African states, Britain and France still succeeded to have continued leverage over their respective territories politically as well as economically. The intelligence of the predators was once more luring the prey into the fangs of the enemy, the consequences have been devastating.

The lack of foresight from the Africans, to know that the very reason that brought Europeans to colonise and exploit them hasn't changed, the creation of organisations has led future generations into more trouble. The same Europeans who fought tooth and nail to hold on to the colonies couldn't in any way have a change of attitude under these organisations. The pride of lions had once more triumphed over a herd of muscular and powerful, yet very stupid buffaloes. As I read about Julius Nyerere, Kwame Nkrumah, Thomas Sankara, the few African leaders who had warned of the dangers of individualism, it sickens me knowing what Africans could have achieved for the continent if they came together.

The European colonialists promised Africa independence, they made the entire world believe that "we are giving these people their deserved independence" and the world believed them, even some from the continent of Africa believed. With much pomp, many congregated to honour these so-called "Independence Day" dates with joy. Unfortunately, they are simply acting like those whose goats have been stolen by their neighbors, but who keep holding to the leash after promises of returning the goats.

As I watched a buffalo being suffocated to death by a few lions, another buffalo from the herd mustered the courage to chase the lions from its brother but was met with stern eyes and menacing roars from the lions. Surely the lions' roar told him "If you do not mind your business and backup, the fate of your friend will be yours" The only courageous buffalo of course withdrew because he didn't have backup from his herd. I have watched other films showing a buffalo in captivity and its crew cautiously edging closer to the captured one and, suddenly out of fear of the buffaloes, the few lions run away in fear of the buffaloes. The rare courage shown by buffaloes always brought a smile to my face. "At least I have witnessed the hunter being hunted" I say with some satisfaction.

The question of intelligence still intrigues me; "Can one race be more intelligent than another? Like many black men or Africans to whom this question is directed, the answer is obvious. "Absolutely they aren't more intelligent than us" Probably it is race pride that tells us we aren't any less intelligent than the 'Whiteman' while at the same time acting with the stupidity of the wildebeests and buffaloes that let themselves fall prey to a few lions. Maybe these powerful animals also argue among themselves and conclude that the big cats that use them as food are not in any way more intelligent than they are, though the obvious is true.

"Maybe God made the Blackman to be this foolish" I tell myself as I try to avoid thinking of Africa's issues. "The Democratic Republic of the Congo is a vast, yet tormented place. This country is rich, yet its people are some of the world's poorest" This is according to the BBC's 'A Journey into the Heart of Africa' which tells the story of the Belgian annexation, the atrocities and exploitation of the Congo by King Leopold II. Supposed to be one of the richest countries in the world, The Congo rather reels end in poverty. The same scenario characterizes almost all countries on the continent with no exception.-

I once brought up the topic of race and intelligence with my friend from secondary school who now resides in Germany.

"It is high time we accept the facts." He said. "Bro, I had this same discussion with some friend's years ago, but they claim there is nothing like some people being more intelligent" My friend continued "I mean the evidence is clear, I feel bad being referred to as less intelligent. When we look at Africa's standing against the rest of the world, can we even argue this thing of intelligence?" He asked.

"It hurts, but truth be told. Maybe this truth should be told often so that it will spur us to challenge ourselves and wake us up from this slumber" I felt the anger and frustration in Hassan's tone as he spoke. Before we concluded our discussion, we both agreed that intelligence needs to be challenged, put to work and the results will be different. Africans have undeniably let their intelligence sleep while others who had awoken theirs keep trampling on Africa.

Thomas Sankara rejected foreign aid as well as loans from international lending bodies. "He who feeds you controls you" he told his people and encouraged all of them to go to work. "Farmers of Burkina Faso; hoes yesterday, hoes today, machinery tomorrow" he told peasant farmers in 1987 at a locality called Pibaore. The results in crop production were enormous, his country went from food dependency from Western donors to self-sufficiency.[9]

CHAPTER 5

THE ARRIVAL OF THE SCAVENGERS

CHAPTER 5
THE ARRIVAL OF THE SCAVENGERS

A coalition of cheetahs lays low in the bushes savoring their meal; they just killed an antelope and as usual are enjoying the fruit of their intelligence. The cheetah, they say, is the fastest hunter in the wild. Barely have they consumed half of their meal than some white specks are spotted from the horizon above the Savannah grasses. They are heading towards the direction where the cheetahs were having their feast.

"Ahh, this is some trouble here for them". The film narrator explained.

A pack of wild dogs noticed there was a catch somewhere and they set out to take advantage of it. "The African wild dog uses its sense of sight to find its prey, not its sense of smell. They also do not use cover tactics when approaching their prey. They can run for up to 66 km/h (41 MPH) for several kilometers. Fast and ferocious, they excel at hunting the speedy" the National Geographic Wild film narrator narrated.

Before the feasting cheetahs knew it, the wild dogs had circled them, they outnumbered the cheetahs about four to one. The wild dogs knew they had an opportunity to have a meal, howling around the feasting cheetahs to distract them. The narrator stated that wild dogs, like hyenas and vultures are experts at chasing the big cats away from their catch. "They are resilient and annoying. I think sometimes they just annoy the big guys and they cannot keep up with their annoyance." the narrator went on.

A few cheetahs tried scaring the intruding wild dogs away, while some ran away as the cheetahs charged at them, others tried pulling the meal away but then the cheetahs made a quick turn to scare these other wild dogs. A most often used strategy by scavengers to distract bigger animals from all angles. Finally, the coalition of cheetahs realized they couldn't keep up, so they gave up on their meal and retreated while the dogs took over and tore apart the antelope voraciously.

I disliked the scavengers but I have come to learn a thing or two from their savage attitude. Though small in size, their power of togetherness is a force to reckon with in the wild. They don't easily back down, not even to the king of the jungle. They move together like a stream, and when one is attacked by an adversary, they all fall on the adversary as one. They have a "One for all, all for one" mentality which is indispensable in a world where survival lies in every bit of strength and intelligence. I have watched in amazement how wild dogs push the mighty "king of the jungle", the lion, to a tight end, fighting for its life against the wild dogs. It pleased me to see the powerful lion, fighting not for food but for a chance to live.

Growing up in Cameroon, we did not live anywhere close to the river that the Portuguese named Rio dos Camarões (River of Prawns) when they first docked at the then coast of Cameroon on their pioneer voyage across the West African coast. However, we always had a lot of dried seafood because our father was a driver who worked most of the time in the coastal region. Apart from the fish, he also brought back items like coconuts, sugarcane and other fruits that were very scarce in our part of the country or came solely from the coast.

It was around class five in primary school that we started learning the history of Cameroon. When the Portuguese explorer Fernando-Po arrived at the coast, he noticed lots of prawns in the river and gave its name Rio dos Camaroes (River of Prawns), our history lessons from primary right up to secondary school taught us. Back home from school, we all shared what we learnt at school and Dad in turn told us stories of how vast the sea was, how leveled the topography of the coast was.

"You look into the water from the shore, and you see no end, it's just water going endlessly." Dad always told us of the vastness of the waters. Such stories about the coast made us long to one day visit and experience this for ourselves as all we were used to seeing were small streams and rivers.

The vast waters along the coastline of West Africa have for centuries been a source of sustainability for its various indigenous people. I have learnt that at particular seasons of the year, the coastal people of Cameroon perform sacrifices to the gods of the sea to provide them with lots of fish, a tradition which other people along the West African coast perform as well. Just as these indigenous people from the coast beseech the gods of the sea for more fish, so too other tribes in the hinterlands perform some rites at the beginning of every farming season for the gods of the land to grant them a good harvest from their farming.

In recent years though, it appears the gods that provide these coastal people of Africa with a great catch have for some reason or reasons refused to grant their requests anymore. Fishermen in most of Africa's waters have complained about the reduction in the number of fish in their waters.

In 2008, I visited the town of Limbe, a coastal town known to many as the hub of fishing activities in Cameroon. My expectations about the dwindling amounts of fish in this part of the West African coast were confirmed. Mokwele, a friend of mine and native of Limbe from Bimbia, was so frustrated with the reality they had come to accept.

"Growing up in this village, Bimbia, we used to go by the banks of the water, cast our nets or place our hooks and catch so much fish that we even returned some back to the water" Mokwele recounted with a tinge of anger. "Today, the waters are so empty, even the small fishes that we used to return to the water a few years ago, we rarely have them today. These days, the fishermen venture into the deep ocean and still their catch is nowhere near what they used to catch. Some of these fishermen have abandoned fishing as riding okada (motorbikes) has become more productive than fishing these days." Mokwele's explanations seemed endless.

"But what really accounts for the fall in the number of fish in the waters" My question to Mokwele was rather a glib one since I knew very well about the fall in ocean fish along the coast of Africa.

"The Chinese bro" Mokwele said, disappointment written all over his face.

"They fish from the deep seas to the shallow waters; their nets are more like sieves that pick up every little fish from the waters. You know our fishermen never used to catch the smaller fish. But the Chinese literally sweep the water and leave nothing for tomorrow."

It was not the first time I was getting this complaint.

The Cameroon state media reported that the Chinese vessels are permitted to fish in the deep waters as the shallow waters are for local fishermen, but other media such as Aljazeera, has reported how Chinese vessels fish in Africa, not respecting any limits. Worst is the type of fishing nets they use which not only drain the African waters of fish, they literally sweep even the seaweed thereby making it difficult for fish to breed. Their use of industrial vessels against small boats used by the African fishermen greatly disadvantages the Africans. African fishermen testified in the Aljazeera report how the Chinese in the bigger vessels sometimes harass them in the deep waters.[1] The Chinese excursions into West Africa come as Beijing has started trying to better regulate its own fishing policy, according to the Financial Times.

"While the Chinese government is starting to eliminate some of the most destructive fishing practices in its own waters, the loopholes in existing policies lead to a double standard in Africa," Ahmed Diame, a Greenpeace Africa ocean campaigner, said in a statement published on the Greenpeace website.

The cases were reported by the Surveillance Operations Coordination Unit of the Dakar-based Sub-Regional Fisheries Commission, various national lists of infractions and by Greenpeace itself, the study said.

The absence of efficient fisheries management in some West African states allows rogue companies to plunder marine resources, the BBC's reports in May of 2015. In less than a month, Greenpeace documented an average of one new case of illegal practice by a Chinese-owned boat every two days, but the report's authors say they think that is only the "tip of the iceberg." [2]

In 2015, Greenpeace International published a report which was the result of an investigation that had been going on for about two years. In the report, they say at least seventy four fishing vessels owned and operated by four Chinese Distant Water Fishing Companies have been exposed for fishing illegally in prohibited fishing grounds in West Africa and falsifying their gross tonnage.

The Greenpeace investigation implicates the vessels in 82 cases of demonstrated and potential cases of illegal, unreported and unregulated fishing activities and gross tonnage fraud, in Senegal, Guinea, Guinea-Bissau and Ghana, and includes China's largest company, China National Fisheries Corporation (CNFC). The cases date between 2000 - 2014, with the oldest case as far back as 1988.[3]

he China National Fisheries Corporation (CNFC) under-reported gross tonnage for forty four of its fifty nine vessels operating in West Africa, the report alleged, a practice that enables companies to evade licensing fees and could potentially mean they were fishing in prohibited areas.

In 2017, the Fu Yuan Yu fleet was reported fishing illegally in protected areas in Djibouti, Quatz Africa reported.

The Chinese ships were "taking advantage of corrupt officials, weak enforcement and supervision from local and Chinese authorities to the detriment of local fishermen and the environment," said Rashid Kang, head of Greenpeace East Asia's China ocean campaign.

"These guys who fish these waters in Limbe are Cameroonians, Ghanaians and Nigerians. All they've ever known in life is fishing, but as we speak, they are gradually losing their source of livelihood and life is becoming more difficult for them by the day." Mokwele told me as we continued our discussion.

"It is really sad," was all I could muster.

"The worst thing is that it appears the government doesn't care about what the Chinese are doing, the local fishermen have been complaining to our local authorities to no avail

In Senegal, the fisheries minister issued fishing licenses to vessels of a Chinese industrial fleet involved in illegal, unreported and unregulated fishing activity in recent years, reported by Greenpeace.

The rise of China has been greatly felt on the African continent as it is the case with the rest of the world. It has had the most devastating consequences on the local economies. Apart from Chinese made goods, construction has also become a thing with most projects, if not all being carried out by them. The Chinese have literally taken over the continent of Africa, what some schools of thought have termed the Chinese colonization of Africa.[4] Just as my friend from Germany feared the Chinese takeover of the continent, sentiments about the Chinese 0presence on the continent remain the same, Africans dislike the increasing presence and activities of the Chinese but like the prey that does not know how to defend itself from a pack of wild dogs, it is bound to be devoured by the more intelligent ones.

The takeover of the continent of Africa is happening at an alarming rate, reports say. China's presence is visible all over Africa. But nowhere as much as in Zambia, the African nation in which it invested the most money in 2020. The ties between Beijing and Lusaka are strong and have existed for decades. Today, China possesses one third of Zambia's national debt. China has invested in the mining and industrial sectors, but also in Agriculture. Some Zambians denounce this Chinese presence as a form of neo-colonialism.[5] "These Chinese are so crooked and corrupt; they have no room for human rights and African dictators are just a perfect fit for Chinese policies." My friend Ayong said, sounding frustrated. We talked so much about the aggressive Chinese presence in Africa.

"China is a fast-growing cancer in Africa coming in as a friend, if not tackled now, what they will do to us will be worse than what the Europeans did." Ayong gave me his honest assessment of the Chinese influx in Africa.

Dictatorships that have over the years drowned economies in Africa have been a good breeding place for China to thrive. Dictatorship and corruption are noted to go hand in glove, with Africa having lots of such dictators.

The newly elected president of Tanzania, John Pombe Joseph Magufuli in 2015 was quick to cancel an agreement which his predecessor, Jakaya Kikwete had signed with China. A project worth ten billion US dollars to construct the Bagamoyo port in Tanzania which was in line with China's major connectivity initiative in East Africa involving roads and railways. Magufuli and his administration saw many flaws in the signed agreement which was to give China rights over the port for a century. No other port was to be constructed along the coastal stretch once they built the said port.

"Their terms were exploitative," Magufuli told reporters. In addition, "the $50-million given out to compensate those displaced by the new port project did not reach the beneficiaries in Bagamoyo but was diverted to benefit a few individuals in Dar es Salaam." Magufuli claimed. [6]

Bribery and corruption have worked so well for the Chinese who come to Africa for business and other activities that Aljazeera in its report on Chinese vessels crossing their bounds from the deep sea to shallow waters at the disadvantage of the local fishermen says. "These Chinese fishermen, however, have their protection from the top officials" The same was echoed by the men along the coast of Cameroon who said government officials received money from these richer Chinese people and allowed them to areas allocated for them, the locals. Also, when marine guards intercept Chinese vessels encroaching into low water areas, the marine guards simply collect money from the Chinese and leave. These guards however tend to be very hostile when local fishermen venture into deeper waters.

The scavengers (Chinese) have just found a huge animal (Africa) that has been paralyzed by Western policies for centuries, where nepotism and corruption are the order of the day. They are now wrestling with the bigger predators on who will own a greater portion of the prey.

Africa's politicians simply receive money from foreign investors and turn a blind eye to the plight of the people, the reason why more and more Chinese are finding their way into Africa at an alarming rate. Large farmlands in Mozambique are reportedly in Chinese hands, Zambia fast becoming owned by China as the former is not able to pay its huge debts and the latter is requesting Zambian mines as collateral for more loans. Zambia is not the only one in this basket, the rest of the continent has been held captive by the scavengers and their sharp teeth are busy tearing off the flesh from all angles.

Prof PLO Lumumba, one of Africa's finest brains of this generation in his famous talk held in Dar es Salaam, Tanzania quoted that the Chinese in South Africa celebrated by toasting glasses at the victory of the ANC at their national elections. "I won't tell you why they celebrated this win so much, but you know as I do" Lumumba closed his talk with the Chinese celebrations in South Africa. The reasons of course needed no more explanation. If the ANC had lost, probably the fate that met the Chinese in Tanzania would have met them in South Africa as well. Loan conditions from China have no "eyes" as the Western powers sometimes deny loans citing human rights violations and even corruption. All corrupt leaders from across the continent therefore rush to China, and in no way will China want to see such leaders go.

Watching African leaders get ushered in during the China - Africa summit of September 2018, I saw a bunch of chickens being lured in for eventual capture. "Chicken number one, number two, number three …" I said to myself, counting from the first president right to the last greeted by the Chinese premier. This bunch of chickens, some so old that walking is such a challenging endeavor, got on my nerves. I was furious how some of these people, though having ruled for over three decades, had not understood that this guy offering them bait (loans) is simply playing them into his trap. I was even more furious to notice some young, some newly elected leaders falling for the same trick. Their predecessors fell for such deceit and the current generation still suffers from their foolishness.

On the 25th of February 2021, international news channels repeatedly broadcast the Chinese president, Xi Jinping's message in which he declared that China has secured a "complete victory" in its fight against poverty. According to Xinhua.net.com, Xi made the announcement while addressing a gathering held in the Great Hall of the People in Beijing to mark the nation's poverty alleviation accomplishments and honor model poverty fighters.

"At what cost has China removed her people from poverty?" I asked this question as I digested the pronouncements made by the Chinese president. Of course, it doesn't matter what it takes the predators to satisfy themselves and feed their offspring, the scavengers aren't any different from the big cats in accomplishing their mission. Just like the French and the other western nations that have fattened themselves from Africa, China too is following suit.

When PLO Lumumba quotes "he who feeds you owns you" he simply quotes today what Thomas Sankara had stated as far back as the 1980s. Decades after, the whole leadership of Africa still cannot understand the perils of dependence. "It is not the scavengers' fault that they can manipulate a bunch of healthy and muscular buffaloes, it is the stupidity of these buffaloes that cannot understand their strength and use it for their good." I muttered these words to myself as we continue to dig ourselves deeper and deeper into debt. Maybe Dr. James Watson was right after all in saying we are not as intelligent as the white race. The only thing that disproves this man's statement has been our pride, otherwise there is nothing to show for our own supposed intelligence.

CHAPTER 6

RELIGION

CHAPTER 6

RELIGION

All of the history books we studied at all levels in school showed that European missionaries introduced Christianity to the continent of Africa. This narrative is however not entirely true because Christianity had been practiced in parts of the continent earlier than the coming of the Europeans. In places like Ethiopia and Egypt, Christianity had been practiced long before the European conquest of Africa.

The history of Christianity in Egypt dates back to the beginnings of Christianity itself. Many findings hold that Christianity was brought to Egypt by the Apostle Saint Mark in the early part of the mid-first century according to Coptic tradition and teachings.

Saint Mark's first convert in Alexandria, Egypt was Anianus, a shoemaker who later was consecrated a bishop and who became

Patriarch of Alexandria after Saint Mark's martyrdom in 68 AD. [1] In Ethiopia, Christianity was introduced to Ethiopia in the 4th century when two Syrian Christians, Frumentius and Aedissius went to Axum and started to tell people about Jesus Christ and the Christian faith. Both missionaries influenced King Azena, ruler of Axum who then declared Christianity as the official religion in his kingdom. The Ethiopian Orthodox Church is one of the oldest organized Christian bodies in the world. [2]

The West introduced the new religion to other parts of Africa when they colonized Africa, most of which still practiced African ways of worship, most of which are still practiced till date.

According to Arab oral tradition, Islam first came to Africa with Muslim refugees fleeing persecution in the Arab peninsula. This was followed by a military invasion some seven years after the death of the prophet Mohammed. Under the command of the Muslim Arab general, Amr Ibn Al-Asi. It quickly spread west from Alexandria in the North of Africa, reducing the Christians to pockets in Egypt, Nubia and Ethiopia. [3]

Christianity and Islam are today the dominant religions in Africa, though the practice of African traditional religions has never been completely abandoned as well. In most parts of the continent, traditional religion is practiced coupled with either one of the adopted religions depending on which one was adopted.

African Traditional Religions share some basic characteristics of religion with others including the fact that ethics and rituals have been passed down through texts and orally for generations. Failure to live by the beliefs or to perform certain rituals always attracts some sort of punishment from the deity, so it is preached. Christianity and Islam, just like Africa's numerous religions all share the concept of discipline when followers fail to meet up to their standards.

In Africa, the mainstream religions have emerged to be Christianity with an estimated 40% of the population, and Islam forming 45%. According to all the major surveys, Christianity and Islam each represent approximately 40 percent of the African population. Christianity is more dominant in the south, while Islam is more dominant in the north. Indigenous African practices tend to be strongest in the central states of Africa, but some form of their practices and beliefs can be found almost anywhere in Africa.

Nevertheless, since 1900, Christians in Africa have grown from approximately seven million to over four hundred and fifty million today. Islam has experienced a similar rapid growth. [4]
These creeds, according to their systems of worship, have elevated a group of people who possess immense power. Imams, pastors and priests are regarded as shepherds of the various flocks who sometimes hold more sway over their flock than heads of states.

In Islam, Imams are considered as spiritual and political successors of the prophet Muhammed who according to the theology of Twelvers, are exemplary individuals who rule over the community with justice, they keep and interpret sharia. Their words and actions, just like those of prophet Muhammed are a guide and model for the community to follow. They are believed to be chosen by divine decree through the prophet.

In the Roman Catholic Church, a bishop is an ordained minister who holds the fullness of the sacrament of holy orders and is responsible for teaching doctrine, governing Catholics in his jurisdiction, sanctifying the world and representing the Church. Other churches like the Ethiopian Orthodox Church, the Coptic Orthodox Church of Egypt, appoint their leaders following the traditions of the churches. All the clerics command a great deal of respect from their followers, they all are viewed as God's messengers on earth and their utterances have power to unleash unimaginable consequences.

Some schools of thought suggest that Africa's numerous problems can largely be attributed to the two major religions: Christianity and Islam. "Religion can be used as a weapon to foster peace or destroy it" as advanced by a German magazine, DW, which goes on to say, "The person doing the interpretation can present religion to a people as a tool for their good or bad." In a Deutsche Welle report of 29th June 2021, people from across Africa attest that besides its positive impact, religion has undoubtedly contributed greatly in the continent's numerous problems. (5)- https://www.dw.com/en/religion-and-development-in-africa-blessing-or-curse

As such, the people who argue that religion has helped in relegating Africa hold the view that religion has been used as a weapon against Africans, a statement with which I agree greatly. Rev Walter Mwambazi during a Ted Talk held in Lusaka on October 8th, 2020 states that "Religion has been historically used as a means to systematically oppress, suppress and take control of resources by the rich elite from the poor." He advances five reasons why religion is the number one cause of poverty in Africa. [5]

When King Leopold of Belgium penned a letter instructing his fellow missionaries on how to present the gospel to the colonized people of Africa, it was intended as a weapon of oppression against the natives. Thus, a majority of Western missionaries to Africa never preached a gospel of salvation or self-assertion, rather they showed how good colonialism was to the people, covering up its many horrid and dehumanizing aspects. These Christian missionaries were another tool used by their respective countries to make sure that the Africans remained in submission to the colonial authorities.

It was the missionary bodies that encouraged native Africans to adopt European names upon converting into Christianity or enrolling into the mission schools. Most, if not all schools in the colonial era were run by missionaries, and they indoctrinated Africans to adopt Western names.

The iconic anti-apartheid activist and former South African president, Nelson Mandela revealed how he adopted the name 'Nelson'. "No one in my family had ever attended school. On the first day of school, my teacher, Miss Mdingane gave each of us an English name. This was the custom among Africans in those days and was undoubtedly due to the British bias of our education. That day, Miss Mdingane told me that my new name was Nelson. Why this particular name I have no idea" Nelson Mandela revealed in his 1994 memoir. 'Long walk To Freedom'.[6]

The former Ghanaian president, Jerry Rawlings also explained how he opposed the missionaries' ways of imposing their culture, saying. "Christianize me if you may, but don't try to Europeanise me" during his interview with Italian journalist, Flavio Marinoni.[7] The same practice was followed in areas under Islamic influence where Arabic names have mostly replaced indigenous African names.

Colonialism from its inception was intended to exert the power of the 'white race' over the people of Africa, a process which was achieved through force. Then the people were coerced to work for the masters with little or no pay. Depending on which part they carved out for themselves, the resources of the marked territory belonged to the various colonial masters, a practice which has persisted till date. It is sad to note that all these abuses went on under the watchful eyes of the Church. As a moral body responsible for promoting societal justice, the missionaries never called out their governments against such ills or encouraged the converted natives to avoid exploitation by their fellow European brothers, they enabled it rather.

David Livingstone, one of the most popular missionary explorers of the 19th century fought hard to end the East African - Arab slave trade. He was an anti-slavery crusader but at the same time an advocate of British commercial and colonial expansion in Africa. Livingstone died in Africa, in present day Zambia while still on his pursuit to discover the interior of Africa for his country. His contributions in opening up Africa for his country earned him fame and respect in the United Kingdom.

Today, there is great fear that religion is in no way advancing or tackling the many problems plaguing Africa. The socio-political problems of Africa have been ongoing since Africa demanded for independence from the Western colonial masters. The Clerics, whom the people look up to as messengers of God have had little or no contribution in redressing their myriad challenges. The Bible says that all wrongdoing can be traced to an excessive attachment to material wealth.

"For the love of money is the root of all evil: which while some coveted after, they have erred from the faith, and pierced themselves through with many sorrows." 1 Timothy 6:10 (KJV)

Perhaps, the clergymen/women have been caught up in the very trap that the Apostle Paul warned against- money. Maybe the so-called men and women of God have largely focused on wealth rather than molding the very people they were supposed to.

In East Africa, the Maasai herdsmen have the responsibility to lead their livestock to the best fields. "There is no Maasai without land" Kimani Maruge explained in the movie "The First Grader", the film based on the true story of a Maasai man who fought British colonial rule with the Mao Mao revolution which contributed to Kenya's independence from Britain. They needed their land for grazing, a deep-rooted inheritance of the Maasais. This story is true of grazing land all over the continent. Scrawny looking cattle imply malnourishment and indicate that the shepherd has not been doing a good job with the herd.

The word "pastor" is derived from the Latin noun pastor which means "shepherd" and is derived from the verb pascere – "to lead to pasture, set to grazing, cause to eat". The term "pastor" also relates to the role of elder within the New Testament and is synonymous with the biblical understanding of minister.

In Islam, such shepherds or guides as in Christianity do exist to lead their followers on the right path, the path of a good fulfilling life.

In the Shi'a context, an imam is not only presented as the man of God par excellence, but as participating fully in the names, attributes, and acts that theology usually reserves for God alone. Twelvers and Ismaelis Shi'a believe that these Imams are chosen by God to be perfect examples for the faithful and to lead all humanity in all aspects of life. They also believe that all the imams chosen are free from committing any sin. These leaders must be followed since they are appointed by God.

According to Jacob Olupona, professor of indigenous African religions at Harvard Divinity School in an interview with the Harvard Gazette in 2015, revealed that Christianity and Islam in the past century have grown tremendously in Africa. He disclosed that according to all major surveys, Christianity and Islam each represent about 40 percent of the African population. [8]

In a relatively short time, Africa has gone from having a majority of followers of indigenous, traditional religions, to being predominantly a continent of Christians and Muslims. New data from the Gordon Conwell Theological Seminary shows that, for the first time ever, more number of Christians live in Africa than on any other single continent. [9] If a majority of Africans fall under the main religions of the world that preach virtues of life by paragons, yet Africa is in a constant decay. Is it evident that these paragons have failed the very people they have been molding? If the grazing fields are full of grass but the cattle rather depreciate year after year, it may be concluded that the shepherds have been found wanting.

"I am in no way defending these Western stars." My friend Ayong Hassan from Germany remarked during our phone discussion. "You remember when Beyonce, Jay Z and most of these American celebrities were labeled as illuminatis?" Ayong asked.

"Sure, I remember" I responded from my end. He was referring to popular videos on social media that accused most American celebrities of sinister occultic allegiance.

"We were made to believe that these guys only succeed because of an extra force, a force of darkness. But I am telling you, these guys work extremely hard to be where they are"

"Sure, they do" I concurred to his statement. "I followed Beyonce's daily routine and to be honest, it is not an easy task putting in as much work as she does. I mean she practices for hours a day and that's what makes her and the others succeed. But we spiritualize everything and don't work, yet expecting a miracle every day."

"Truth be told, we Africans are lazy and religion has greatly impacted that in us. If we need to catch up with the rest of the world, we need to get out of this illusion that we will pray about things, and they will be fine. Hard Work needs to be preached and encouraged rather than sitting and labeling hard working and successful people as illuminati." Ayong poured his heart out as I nodded in agreement.

"There are basically two economies in the world and every nation on the face of the earth falls under either one of them; Knowledge and Religion economies. The rich and prosperous nations of the world belong to the knowledge economy, while the poorest and poverty-stricken nations fall under the religion economy. The knowledge economy is one that engages the brain, asks questions rather than questions solutions to problems. The religious economy is one that blackmails God and dumps any responsibility on him. The knowledge economy tasks the brain while the religion economy numbs the brain. Japan with no natural resources is a knowledge economy and one of the most prosperous on the planet. At least six of every ten cars on Africa's roads today are Japanese made. Japan is not a religious economy, yet it is one of the most corrupt free nations. China is not a religious economy but nations that ignore China, do so at their own peril."

African nations fall under the category of the religious economy. More so than others, African countries are endowed with the majority of the natural resources on the surface of the earth yet also among the poorest. Africa proudly wears the badge of the poverty capital of the world with endless corruption, it is also one of the most religious continents on the planet. God watches in disgust as we mortgage our brains. But here is the irony, the two major religions of the world, Christianity and Islam emanated from Israel and Saudi Arabia respectively, but none of those two nations fall under the religion economy. Now, one of the evidence of their subscription to the knowledge economy is the yearly pilgrimages that generate revenue for them in billions of dollars. But hypocritically, some corrupt nations are comfortable disseminating nuisance to humanity with their globalization of religion. There is a reason he gave man a brain and instructed him to subdue the earth. If you are a Christian or Muslim who does not commit yourself to the pursuit of knowledge, your life will become a perfect reflexion of some of these uninspired nations. It is an illusion to anticipate wisdom when people do not pay attention to knowledge" Anthony Afolo of the New African Broadcasting Network (NABN) analyzed Africa's take on religion. [10]

Anthony Afolo's comment about China and Japan as purely knowledge economies however got me thinking. I did some research and found out his statement was not entirely true. Buddhism and Shintoism are the major religions in Japan. According to the annual statistical research on religion in 2015 by the Agency for Culture Affairs, government of Japan, 70.4 percent of the population practices Shintoism, 69.8 percent Buddhism, 1.5 percent Christianity, and 6.9 percent others.[11]

China as well is not purely a knowledge economy as pointed out. China is a multi-religious country. Taoism, Buddhism, Islam, Protestantism, and Catholicism have all developed into culture-shaping communities throughout Chinese history. One may conclude that countries like China and or Japan mentioned earlier prosper because their shepherds lead them in the right path, that they apply religion in a way that build's the people's capacity to fend for themselves rather than sit and expect religion to open doors for them as it is the case with Africa.

Therefore, it may be concluded that Africa's shepherds have failed in their duty to lead their flocks on a rightful path, not using religion in its rightful context to build the flock's intellect.

Some argue that it is not religion that has held Africa back, it is rather the application of religion which is part of a complex reason for Africa to remain subdued. Some clerics have condemned their people for pursuing education as they termed it evil. These same people however have never stopped using machines born out of education to promote their religious beliefs. Time and time again I've watched some of these religious leaders use social media to pass their harmful messages, they continue till date to use these machines to advance their beliefs even forcefully.

My friend, Njoh, was one of the most brilliant brains while we were in secondary school. He excelled so well in all subjects at the same time that we all were equally involved in Christian activities in school. I once asked him if pursuing studies in sciences caused him to think less about God.

"No, not at all" Njoh responded without any second thought. "I think my studies in sciences from secondary school till now rather make me appreciate God more because when you study all these magnificent works of his, then you even have great reverence for him" Njoh gave me his take on science and its relation to God. "There is absolutely nothing created by man which is not a combination of what's already been in existence. So those people who reject science on the basis of it undermining who God is are completely naive in their thinking" I couldn't agree less to my friend's remarks.

In reading about some great scientists who changed the world and mankind, I was in awe to learn that some of them were preoccupied with religious activities as well, given that they identified themselves as Christians. Isaac Barrow was an English Christian theologian and mathematician who is generally given credit for his early role in the development of infinitesimal calculus; in particular, for the discovery of the fundamental theorem of calculus. His work centered on the properties of the tangent. Barrow was the first to calculate the tangents of the kappa curve.

Michael Faraday was a devout Christian; his Sandemanian denomination was an offshoot of the Church of Scotland. Faraday served the final two years of his second term as elder prior to his resignation from that post. Biographers have noted that "a strong sense of the unity of God and nature pervaded Faraday's life and work."

Sir Robert Lewis Fullarton Boyd is another scientist who had a passion for God and his field of study and work. He was a pioneer of British space science and founding director of the Mullard Space Science Laboratory (part of University College London). Boyd was a committed Christian and he saw no conflict but subtle complementarity between science and religion. His 1950s and 1960s writings and lectures on this relationship helped Christians to be more influential in science, through the Research Scientists' Christian Fellowship.

Benjamin Solomon Carson Sr. popularly known as Ben Carson is one man whose achievements are greatly appreciated in today's world. He is considered a pioneer in the field of neurosurgery. He became director of Pediatric Neurosurgery at the John Hopkins Children's Center in 1984 at age 33 as the youngest chief of pediatric neurosurgery in the United States. His groundbreaking achievements in neurosurgery earned him national and international recognition. Of interest is the fact that Ben Carson always attributes his success with God and his publications rather challenge the mind to unleash its potential and not rely on God to solve its numerous challenges.[12] Ben Carson titled one of his books "Think big, unlashing your potentials for excellence. In 1992, he stated in his autobiography that his personal philosophies of success and what he sees as the stabilizing influence of religion.

CNN's inside the Middle East in 2010 revealed lots of Muslim inventions that shaped the modern world, discoveries from all works of life which have contributed immensely to human wellbeing.[13]

It is the rhetoric that stems out of some predominantly Muslim countries in Africa that will go on to validate the notion of religion being used as a weapon against Africa.

In Sudan, Islamic schools known as "khalwas" teach children to memorise the Koran have existed in Sudan for hundreds of years. They are tens of thousands of them in such schools, children as young as five years old are given to "Cheiks" who run the schools where the children are groomed to recite the Koran.

BBC's Fateh Al-Rahman Al-Hamdani in 2018 reported in an undercover report where children who are supposed to study the basic things that contribute to their wellbeing rather spend their years reciting the holy book. [14]

The notorious Islamic terror group 'Boko Haram' that stemmed from Northern Nigeria and which has so far spread to Northern Cameroon and other countries in the area is said to be influenced by the aspect of education.

According to the Oxford Learner's dictionary, the word "haram" in Arabic means forbidden or not permitted by Islamic law. Boko haram meaning Western education is forbidden means children in such areas most definitely grow up lacking behind as education for them is a taboo.

What they use to pass their message is technology born from the same education they so much detest; the weapon which has been used for centuries to hold the people back being the religion which supposedly was meant to set them free.

CHAPTER 7

GHOSTS OF LOW SELF-ESTEEM

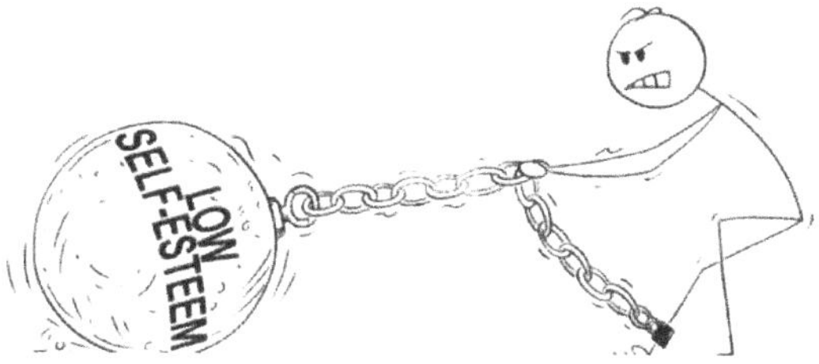

CHAPTER 7

GHOSTS OF LOW SELF-ESTEEM

Stories of ghosts were common in our growing up years. These ghosts, we were told, roamed around in certain areas in the village during particular hours of the day. Whether or not these ghosts ever existed remains a mystery. But while we were young, we dreaded them, avoiding such places during hours of the day in the village when ghosts were alleged to be active.

We eventually outgrew our fear of ghosts, still, it amazes me how the notion of something unseen can easily instill fear. Take predators in the wild who need merely to roar, to send their prey fleeing into the bushes for example. The minds of humans are easily susceptible to fear.

Although humans may seem different from other animals, they are remarkably similar-, from mice to monkeys, animals have similarities. They only differ in cognitive abilities, such as a full-blown language as well as reasoning and planning abilities.

Humans have succeeded in capturing the minds of all types of animals. They have also triumphed in subduing other humans to be answerable to them. When the mind fails to think rationally, it fails to ask questions as to why things are the way they are. Unlike the animal's mind which doesn't ask questions (according to Quora) the human mind does, it demands for changes in areas where the human being doesn't feel comfortable. This is a fundamental aspect of humans that rulers have tried to suppress, particularly because it involves challenging their authority.

I once Listened to Dr. Martin Luther King Jr.'s last famous sermon delivered at the Ebenezer Baptist Church on January 4th, 1968. Referred to as "The Drum Major Instinct." In it, he narrates the tragic prejudice, from which one can understand how one race perceives and places itself above another. Says Dr King,

"Do you know that a lot of race problems grow out of the drum major instinct? A need that some people have to feel superior. A need that some people have to feel that they are first, and to feel that their white skin ordains them to be first. And they have said over and over again in ways that we see with our own eyes. In fact, not too long ago, a man down in Mississippi said that God was a member of the White Citizens Council. And so, God being the charter member means that everybody who's in that has a kind of divinity, a kind of superiority. And think of what has happened in history as a result of this perverted use of the drum major instinct. It has led to the most tragic prejudice, the most tragic expression of man's inhumanity to man." An excerpt from the "Drum Major Instinct" sermon delivered by Dr. Martin Luther King Jr.[11]

In school, we studied the "Age of Enlightenment" in European history, the period when people challenged despots, questioned the teachings of the church and even the existence of God. It was so strange for us to discover that most, if not all European countries used to be ruled by absolute monarchies. The Age of Enlightenment that started in the 17th century was a major turning point for Europe as monarchs abdicated at the demand of the masses. The ideas of the Enlightenment undermined the authority of the monarchy and the Church and paved the way for the political revolutions of the 18th and 19th centuries. These ideas shaped history, thanks to the works of people like Adam Smith, Bacon and Descartes, Beccaria, Rousseau amongst others whose works did not just end in Europe but helped shape the American revolution as well.

After centuries of living under the absurdity of inferiority and unable to challenge the monarchies, the era of "thinking" shattered the notion of acceptance from the ruling elites. Everyone could now aspire to become a leader, their rights were guaranteed, and freedoms gained . Things which had long been reserved for a particular class of people, whether merited or not, could be accessible to everyone. Leaders who failed to meet up with the people's expectations were replaced with others, decisions were made by the people through universal suffrage.

In the late 18th century, as the price of bread steadily rose amidst other factors in France, the French people had no other choice than dethroning and subsequently killing their monarch, Louis XVI. Similarly in Russia, inflation, poverty, hunger and other factors forced the Tsar to flee when the Bolsheviks led a popular uprising. It was the same with other monarchies in Europe who completely gave up power or designated more power to the people, the case of the British monarchy.

It was the same spirit of thinking differently from what the masters had educated the colonies about that pushed great Africans to demand changes in relations to colonialism. Africans from the north to the south questioned and challenged European hegemony over them. They put up a fierce resistance which eventually led to the decolonization of the continent, but unfortunately it was a limited decolonization. From Houari Boumediene of Algeria to Jomo Kenyatta of Kenya, Sékou Touré of Guinea, to Samora Machel of Mozambique among all other anti-colonialists, these brave African leaders refused to continue in servitude to the Europeans.

Colonialism is a system that has had deep far-reaching effects on Africans. In order to keep the people under their control, the Westerners ensured the Africans knew they were not equal to them. Perhaps, King Leopold II's letter had an impact even on the status of the whites against the blacks, the 'drum major instinct' explained by Rev. Dr. MLK Jr. The very fact that one race captured another and forced them to work as slaves explains the thinking of one seeing themselves as superior to the other and vice versa.

Julius Malema, one-time youth leader of the African National Congress (ANC) in South Africa and current opposition party leader in parliament challenged a fellow white South African. For Malema, "racism is taught and not born with," just as black inferiority has been passed from generation to generation, a position regrettably accepted by Africans and their descendants. While whites continue to assert themselves as superior through generations, Africans on the other end have settled in their spot as inferior through these same generations. Everything 'African' or 'black' is always deemed bad, everything except the rewards that the white man gets from Africa and her descendants. We have been made to believe that every good thing that exists in Africa today is thanks to the white man's coming to the continent, never recognising any contributions that have been made by Africans over the years.

"There is no place for [the Bantu] in the European community above the level of certain forms of labour ... What is the use of teaching the Bantu child mathematics when it cannot use it in practice? That is quite absurd. Education must train people in accordance with their opportunities in life, according to the sphere in which they live." This is one of Hendrik Verwoerd's quotes, a Dutch-born South African politician, sociologist and journalist. He underscores the truism that Africans continually settle for crumbs from their own bread.

Europeans branded Africa, we learnt, "The Dark Continent". With such an idea implanted in the African child's mind early in schools, the Whiteman tends to be revered as a god. With the ability to control almost everything, this belief of white supremacy has been passed down through generations, by whites and blacks alike.

Recently, I followed the BBC's "History of Africa" documentary, a story told by Africans. In the film, Prof. Ngwabi Bhebe of Midlands

State University in Zimbabwe confessed that he had never known Africans' contributions in molding their societies before the coming of the white man. A degree of pride arose in me when I learned that Africa's civilisations had actually been going on before the coming of Europeans. Archaeological studies have shown how Africa had advanced in some areas before such civilisations were destroyed from European conquests and subsequent colonisation. Powerful kingdoms from the southern part of Africa as well as the North, East and West of Africa had been built over centuries as presented by BBC's History of Africa.

The intent of the colonialists had been to portray the white race to Africans as a messiah who came to deliver them from darkness (African way of life) into the light (European way of life). Even the missionaries had a duty to uphold this teaching to their African converts. It was such teachings that triggered President Jerry Rawlings' statement;' "...you may Christianise me but do not Europeanise me."[2]

European names were given to natives at conversion in the case of converting into Christianity, Middle eastern names accepted upon converting into Islam, a practice which has successfully degraded native names. Calling Africans who fell for this doctrine by the tribal name is tantamount to disrespect. "Rejecting my name is more like me rejecting who I am, my identity, rejecting where I come from and my people as well." This is exactly what the Europeans had wanted Africans to do, denying themselves and opting to be like Europeans instead.

On the streets of my hometown, Bamenda and in Africa as a whole, it is common to hear comments such as:

"You know I pay like a Whiteman",

"She's a white woman in a black body",

"You dress like a white", to name these few comments which portray the white race as glorious and perfect people while the black are backward and dejected. In effect, anything bad is deemed black while good deeds or actions are referred to as white. A child who excels in school is referred to at home as white while the one who performs poorly is black, sadly so by the parents even.

"The Whiteman's witchcraft is modeled to help mankind while Africa's witchcraft is that which seeks to eliminate fellow Africans", these we have often heard of when praising the Whiteman for technological advancements or condemning a fellow African for some mystical happening.

It is a ghost which has been present on the continent from time immemorial, a ghost which still goes around reminding the Africans who the Whiteman actually is and reminding the Blackman of his place on this jungle called earth. Even when sons and daughters of Africa challenged the Whiteman's colonial policies and demanded for independence, the Africans could not break loose from the grip of the masters and as a result have forever lived under the influence of the masters that their predecessors fought against.

As the ghost of low self-esteem constantly roams around, reminding us that solutions to present challenges can only come from others than ourselves, we have comfortably settled for this position while praying to God and hoping that the solution providers (Whiteman) come up with solutions to our plethora of problems.

"While America and China compete to see who will be the first to establish the 5G network, Africans sit watching and cheering like spectators watching a soccer match, waiting who will score first" Prof PLO Lumumba laments on the lack of initiative by Africans.

Recently, as the world was hit by the deadly coronavirus (Covid-19) pandemic, the gaze of Africa as always was set on the Whiteman for solutions. The lone African who dares to seek solutions to some of the problems will not be trusted by fellow Africans simply because of the notion that nothing good can come from Africa, only the superior race can bring such solutions.

African universities have over the decades produced countless engineers in various fields but have very little to show for as our infrastructure has been the work of the Whiteman and of recent years, the Chinese. Africa may have the best technicians but that ghost which constantly reminds us of our place in this jungle makes us go to the one we perceive as superior. We even see these fellow sons and daughters of Africa in great numbers who have left home for the West, excel greatly in different fields, excel because they have been given the opportunity to show what they can offer.

The African continent and her descendants have had tremendous success in sports as a whole and soccer in particular. Around the world, coaches of various national teams have always been their retired athletes who in turn get to pilot their various teams. For Africa, the successes recorded by African athletes especially with soccer have also fallen short to be seen in light of training current and upcoming athletes. Europeans have always been and are still regarded as masters in light of training African soccer teams.

Leading up to the African cup of Nations (AFCON) 2019, Senegal had Aliou Cisse as caretaker coach. Cisse was amongst the squad that made history at the 2002 FIFA world cup in Seoul, South Korea with their incredible run that ended at the quarter finals, they were only the second African nation to attain such a milestone. The nation's football federation rejected Aliou Cisse's request to lead his nation as head coach into the African Cup of Nations (AFCON) even with a lower salary compared to the European coaches. "It will be an honour for me to lead my nation as coach," Cisse told his country's football federation when pleading for the job.

The Senegalese federation rather pursued Frenchman Alain Giresse to lead Senegal to the AFCON given his experience in African soccer. They however failed to land the whiteman because he was far too expensive, Aliou Cisse eventually secured the job as Tunisia was able to hire the Frenchman. Subsequently, Aliou Cisse's boys faced Alain Giresse's at the semifinals of the AFCON 2019 with Cisse emerging victorious as Senegal beat Tunisia by one goal to nil and advanced to the finals before losing to Algeria.

Aliou Cisse's story of persuading his fellow countrymen to give him a job he felt he was qualified for but rejected for the ones we have always believed that are architects of the game is typical. His triumph, especially over the one who was chosen over him, has stood out to be one of my best soccer stories of recent times. Maybe if Africa had faith in the very technicians they train in various fields, we would have been better off than sticking to the ones we have come to believe are better than us.

Cisse finally got an opportunity to prove his mettle in the Africa Cup of Nations that was held after the Covid-19 pandemic in Yaounde, Cameroon. Cisse proudly led his Senegalese Lions team to ultimate glory as they brought home the trophy for the first time. He finally proved that an African national can deliver good results, just like the 'whiteman', if not better.

Perplexed if no black had ever made some great discovery in history, my memory quickly reminded me of Dr. Ben Carson whose life, education and breakthrough in medicine inspired the film "Gifted Hands." In 2001, he was named by CNN and TIME magazine as one of the nation's (USA) 20 foremost physicians and scientists and was selected by the American Library of Congress as one of 89 "Living Legends" on its 200th anniversary. Carson's works have not just been inspirational to Americans, they've been to Africans as well. My longtime friend Njoh was partly influenced by Ben Carson's story to study medicine.

Lewis Howard Latimer, who I found out was a former slave who fled from his master to freedom, was an inventor. His invention of the improvement in light bulb manufacturing process was a major breakthrough at the time. Latimer is an inductee of the National

Inventors Hall of Fame for his work on electric filament manufacturing techniques.

George Washington Carver whom I had read about in an Ebony magazine was a son of a slave owned by a German American immigrant, Moses Carver who took the boy as his own and encouraged him with education. His works in agriculture were a major breakthrough as he became one of the most prominent Black scientists of the early 20th century.

The list of African descendants who have made contributions to discoveries thrilled me, but such stories have been relegated to the back and for the sole reason of showing a particular people as superior, the works put on by men and women of African descent, under very difficult circumstances fail to gain recognition. And until now, we have often wondered if actually our ancestors used sharp stones as knives and used just pointed sticks as weapons. The myth shattered with BBC's history of Africa which clearly shows evidence that Africans had made great strides in manufacturing such tools long before the arrival of the Europeans.

Monuments such as the Ethiopian Orthodox Churches at Lalibela that date back as early as the 7th to 13th centuries, the pyramids of Egypt and Sudan, the ruins of great Zimbabwe, ancient process of smelting iron in Burkina Faso, the bronze works of arts by the ancient Benin empire (under Edo state in Nigeria) among numerous works are testaments of who and what Africans had been before the coming of the Europeans as shown by Zeinab Badawi of the BBC.

At the same time that they downplayed Africa's progress and named the continent "black", they were quick in looting Africa's work of arts for themselves. Some of which have been returned while a great amount still fills their museums.

"We don't want or expect all of them to get back. We would even like some of them to be out there, to serve as ambassadors for us because wherever they are seen, somebody will ask where is this wonderful work from? They will say it's from the Benin empire." The Benin Oba, His Majesty Ewuare II responding to BBC's Zeinab Badawi on the topic of repatriating some of the ancient Benin empire bronze artifacts occupying western museums.

Until Africans begin to think deeply, like the men and women during the Age of Enlightenment who saw despotism in Europe overthrown at the demand of the people or as the men and women who stood up and challenged colonialism, we will continue to suffer the same fate for centuries to come.

The world is one big jungle in which survival is for the fittest. All people fight their way through to get their place in this jungle, we Africans have lived under the shadow of inferiority to others, and we cannot continue to live with that mentality any longer if we dream of a better place for ourselves and the generations after us in this jungle. After all, conquered people around the world stood up against the occupiers, most have asserted themselves as equals and have claimed their place in this jungle.

As kids we were told that most Southeast Asian countries used to be on the same economic scale as most African countries, with some even poorer than the African countries.

Most, if not all of these Southeast Asian nations had been captured and held down by the same predators that captured Africa. Today, the stories of these two people are a complete contrast. Nations like China, India and others in Southeast Asia are competing amongst the best economies in the world. Unfortunately, Africans' economies have steadily regressed from where they used to be on the same economic index with their fellow captives of Southeast Asia. We have stayed low ever since we were pulled down by the predators, settled even lower to where they intended us to be. The ghost of low self-esteem still roams around reminding us constantly of our place in this jungle, as such we settle in our low status and pray to God for help or hoping a day will come when the predators decide not to prey on us again.

While we remain in the state in which our captors had wished for us, our only way out is to come together and assert ourselves to where we belong, to where our freedom fighters envisioned for us. And when we do, the predators will no longer tell us who to associate with or do business with. Who we deal with will be the ones who meet our satisfaction the most, not the ones whom we pay allegiance to. Not even the smallest nation will feel threatened anymore because a people of a common ancestry will defend even the weakest of their type from both ends of the Sahara and across the Atlantic.

CHAPTER 8

THE UNPLEASANT PLACE

CHAPTER 8

THE UNPLEASANT PLACE

Early in January of 2018, social and mainstream media went crazy over alleged comments made by the sitting US president, Donald J. Trump. The Washington Post had reported that the president referred to African countries and Haiti along with other Caribbean countries as "shithole countries." An allegation which the president refused in a White House meeting.

Whether the comments were said by him or not, the whole world believed he uttered the words, and they were not taking them lightly. The U.N. human rights office said the comments, if confirmed, were "shocking and shameful" and "racist," while Haiti's foreign minister summoned the U.S. Chargé d'Affaires in Haiti, Robin Diallo for clarification. The African Union claimed it was "alarmed" by Trump's statement. "Given the historical reality of how many Africans arrived in the United States as slaves, this statement flies in the face of all accepted behavior and practice," spokeswoman Ebba

Kalondo said, adding that it was "particularly surprising as the United States of America remains a global example of how migration gave birth to a nation built on strong values of diversity and opportunity."

As the world pounded on Trump over the "shithole" comments, I was however happy that a world leader made such a comment. Trump's comments came amidst discussions over immigration issues, pondering if the diversity lottery (DV) program introduced by President Reggan's presidency and implemented by Bill Clinton's presidency should continue.

Trump had argued that "immigration robs these poor countries of their bright minds rather than do them good" A point which I applauded him for raising given that such programs as the Diversity (DV) Lottery program selects the educated Africans and immigration as a whole drains minds that challenge poor governance thereby leaving the continent in a state of continuous decline. My position of applauding Trump for saying immigration has robbed Africa over the years quickly put me at odds with many members of my community.

Most gatherings in our community during the days following the comments were about weighing in on Trump's latest comments about Africa and other Third World countries.

During Thomas Sankara's short reign as president of Burkina Faso, he once responded to a BBC journalist, Joan Baxter during an interview "Call a cancer what it is, cancer" when she asked if his slogans could be construed as deliberate provocation. His comments were similar to President Trump's, though the world had descended on the US president without mercy because he had simply pointed out the cancer that even the affected person refused to point out.

"The truth is bitter but must be told", is a phrase I have heard most of my life. No one wants to hear the truth, especially when it involves a negative revelation exposing the one concerned. Though a majority of Africans today pray and fast to obtain a visa for greener pastures out of the continent, they do not want their continent to be referred to as such. We want our continent to be showered with the same praises as Europe, America or the advanced parts of Asia, when the truth is that it cannot be compared with these.

The unfortunate truth is, Africa is a "shithole" just as Trump allegedly referred to and her descendants have proven it time and again even without mentioning the word "shithole." African descendants have sought citizenship in other parts of the world in alarming numbers year in and year out. Even when an African works in continental America or Europe, the offspring of such descendants have always been groomed to become citizens of those countries, having very little to do with the motherland.

"My grandfather comes from somewhere in West Africa." Jaokim Noah during his time of play with the Chicago Bulls once said, while answering questions about his ancestry during the "Black History Month" which the National Basketball Association (NBA) strongly promotes.

Joakim's father, Yannick Noah was born in France when his father, Zacharie Noah played professional soccer for a French Team Sedan-

Torcy. Zacharie Noah, Joakim's grandfather, had left Cameroon for France as a result of soccer.

Driving through Tongolo in Yaounde, the capital city of Cameroon, I was pointed at Joakim Noah's grandfather's residence, a well-known figure on the streets of Yaounde who is often seen moving around with his long grey dreads. His son, Yannick Noah brought glory to France and not Cameroon when he won the tennis French open in 1983 after he'd started playing as a kid in Yaounde. Joakim Noah currently holds the French, Swedish and US citizenships, of course with no link to Cameroon for obvious reasons.

Trevor Noah, the South African comedian, political commentator, and television host who is currently host of The Daily Show, the American satirical news program on Comedy Central joked about Africa winning the world cup in 2018. Noah simply joked about what we often do when we watch the French soccer team partake in soccer tournaments. "Here comes Africa united" we will make fun of the French team as it is always composed mainly of the first and second generations of Africans and or people from former French Caribbean colonies in France. The 2018 team's composition was the same as the 1998 team that also won the world cup for France, a majority of the players born of African parents residing in France and some being Africans or from former French colonies who took up French nationality. Noah's comment of "Africa winning the world cup" drew so much resentment from the French people. Noah later read a letter sent to him by the French ambassador to the US during his show, trying to impersonate the ambassador's French accent expressing his discontent with his comments about Africa winning the world cup.

Apparently, France is not the only western country benefiting from African or former colonies' talents. I have taken some time to research the origins of lots of other athletes competing for major European countries and even the USA. From my findings, it appeared that most black players who played and some currently playing for England, Germany, Belgium are people whose parents immigrated to these countries in search of greener pastures. Their parents originated from former colonies in Africa or the Caribbeans with them while some of them were actually born in these European countries. A case in point is the Boateng brothers, born of a German mother and a Ghanaian father, who made different choices. The older brother, Jérôme chose his mother's country Germany while the younger brother, Kevin Prince chose his father's country Ghana.

Athletes in other disciplines have taken up new nationalities or chosen the country born in over those of their parents. Today, names like Anthony Joshua are household names around the world. All through I always thought the big guy was a pure Brit from his name and accent, his ancestral name, Oluwafemi Olaseni told another story of his origins. It was after his victory in 2018 that social media went gaga with celebrations from Nigeria as well. "So, Joshua too is 9ja?" I asked surprisingly after reading comments from the New York based Nigerian blogger, Adeola Fayehun who congratulated her fellow Nigerian brother on his heavyweight victory. Of course, just like others who have chosen the best between two worlds, Africa and the West, official information about him refers to him as "a British professional boxer"

Giannis Antetokounmpo was the first black Greek athlete I came to know. Charles Antetokounmpo, Giannis' father was a former Nigerian soccer player who in 1991, with his wife Veronica moved from Lagos Nigeria to Athens Greece as a part of a wave of immigrants seeking better lives. They left their first son, Francis, behind with their grandparents. Four more sons were born to him and his wife while in Greece. The most famous of them all being Giannis, nicknamed the "Greek Freak" who plays professional basketball with the Milwaukee Bucks and Greece. The Antetokounmpo brothers, three of them are well represented in the National Basketball Association (NBA).

Naomi Osaka who plays top professional tennis was born to a Haitian father and a Japanese mother. Her choice to side with Japan over Haiti in international competitions got people asking questions. It would have been surprising to me if the latter was her decision.

Africans and their descendants have over the years turned their backs on the motherland. It is no secret why all these people and more have chosen all other prestigious and more deserving countries over the less meritorious countries on the continent and the Caribbeans.

I have never heard of any national who is not African, moved to the continent with kids or bore kids while serving in Africa and they later on accepted nationalities for the countries born or residing in. Probably there may be a few though the percentage can never in any way be compared to that which rejects Africa. Africa may hate the shameful 'shithole' label stamped on her, but she is doing little to change it.

If the coin of prestige were overturned, Barack Obama would not have been President of America though his mother was American, Naomi Osaka would have been playing for Haiti and not Japan, the Antetokounmpo brothers and Anthony Joshua would have been flying the Green-White-Green flag of Nigeria at their victory celebrations not Greece and Britain. France wouldn't have won two world cup trophies as well as the other Western countries would not have had African athletes defending their national colours.

A few questions come to mind when I consider President Trump's alleged comments and the fury they engendered.

1. How did we become such repellant people?
2. Who put us in this position that we even get to run away from ourselves?
3. What must we do to make sure that we put to an end this degradation we suffer?

As I ask myself these questions, the answers seem very evident:

1. No, we were not supposed to be the specimen of misery in the world as we have been known for.
2. It is the same people we so often run to who are responsible for the state of shame in which we are, yet we fail to tackle this

monster of a problem as we rather see them as saviors through their imperial policies presented to us.

3. We must see the urgent need to come together and work towards a common goal if we must elevate ourselves to where we aspire to be.

Our breakdowns however cannot always be affiliated with the outside world, we have to an extent been the masters of our own failures.

"Each time I want to fight for African rights, I use only one hand because the other hand is busy trying to keep away Africans who are fighting me." Quotes by Benjamin Burombo of the then Southern Rhodesia, now Zimbabwe. Burombo's frustrations sometimes had come from his own people who rather than join him fight the oppressors, were instead fighting him. Such words by Benjamin Burombo cannot be less true because records have noted that the predators have always passed through fellow Africans to get to their revolutionary leaders. Leader after leader has been betrayed by a very close collaborator who sided with imperialism, handing over the country to the enemies.

The desire to rise to the top is inherent to the human being, unfortunately such desires sometimes mean betraying one's own people. Throughout history, ordinary people have fought their way to become nobles and leaders, just to have brothers or sisters killing their siblings who were leaders to take over their positions even if the process meant collaborating with external help. This is the weakness that the imperial West has exploited to their advantage, aiding fellow Africans who get disgruntled with current leaders to oust them, and in turn making them leaders who remain loyal to imperialists and not the people they rule.

"Africa must find a way to craft her own type of democracy. We must not practice democracy as it is done in the West because we are different. With so many tribes and ethnic groups, our democracy must involve everybody" Prof. PLO Lumumba gave a way forward for Africa to have a better future in one of his many talks.

I'd always listened to FM 105.3 in the hours from 07:30am to 08:00 on my way to work. Steve Harvey and his crew on the morning program "The Steve Harvey show" were my frequent companions when I drove from 1600 Laurel road in Lindenwold through Chews Landing road, through Gibbsboro road to route 73 in Marlton New Jersey where I worked.

I'll wait anxiously for a new prank phone call from Nephew Tommy and the fervent words of wisdom from the one and only Steve Harvey. On this day, America had once more been gripped by yet another shameful act of apparent racism committed on a blackman jogging in his neighborhood. Ahmaud Arbery was killed while doing sports as video footage emerged showing the twenty-five-year-old man during his last moments.

"This country man, we can't take it anymore. We have lived through this our entire lives and from what it seems, they are not going to stop killing us as they have always done. I think the African American population in this country needs outside help. We cannot do this on our own man" My body seemed cold and stiff listening to the words Steve Harvey spoke. His voice told me how saddened and distressed he was concerning the recent killing of yet another unarmed black man in America. I felt the pain from his voice, I felt the frustration and anguish as Harvey mentioned "we need help man, we can't do this by ourselves"

Africa has the moral responsibility to look after her descendants, be they in America, Europe, the Caribbeans or anywhere else in the world. Zimbabwe was punished with economic sanctions because the "white world" never forgot their descendants who went and seized land from Africans. And when the Africans demanded their lands back, the white world rallied and strangled Zimbabwe.

How will Africa take care of her descendants in faraway lands as America when she cannot even take care of her children at home? As yet another unarmed black man, George Floyd who died from Minneapolis cops just a few months after Ahmaud Arbery, Steve Harvey's words again rang an alarm in my ears. Motherland, as I have often heard African Americans describe Africa, needs to play the role of a mother to her children being killed on the streets of America.

Africa was not supposed to be receiving and expecting funds from African Americans or relying on aid as she has been doing since independence. Aid has poured into Africa from countless successful African Americans over the years who because of their ancestry, take Africa at heart. Celebrities like Michael Jackson, Oprah Winfrey, Serena Williams, Stephen Curry, Forest Whitaker and countless others have all been engaged in different ways to bring relief to Africans.

During the 27th World Economic Forum on Africa in 2017, Forest Whitaker talked about his Whitaker Peace and Development Initiative. Just like his peers, his mission involves conflict resolution in some of the war-ravaged places, providing access to education, clean water among lots of social issues.

Africa with her abundant riches was not supposed to be on the receiving end of aid, but imperialism and corruption rather forced Africans to be the people that charity can be measured in terms of who gives more to the Africans.

CHAPTER 9

HEEDLESS LEADERS

CHAPTER 9

HEEDLESS LEADERS

I was a little boy, working with my father on the farm when thoughts began pouring in my mind, thoughts which I knew would be hard to voice out. Then suddenly, as if my brains could bear the pressure of the thoughts no more, I blurted out. "Pa, why can you not sell some of these lands and use the money to better off your life and for our school rather than you work this hard every day while having lands staying unused?" I asked my father. I addressed my father as "Pa", which in my culture is a title of respect given to elderly men, just like the Igbos of Nigeria have "N'nai." Pa is also used for respect given to the successor of an elderly man, in most cases one of the sons of the late father regardless of his age. My father was a "Pa" by inheritance as well as by age.

My question came like a shocker to my father who completely froze in place. Pa Asanji dropped his tools and found a dry spot and sat himself down, then asked me to sit down beside him. I had no doubt that the words he was going to speak were to be taken seriously. "These lands, our fathers fought for them, our fathers preserved them and passed them to us." My dad said while I listened without further questions. He paused in thought before continuing "There is a reason that Baba (grandfather) put me in charge of these lands and not one of my brothers, it is to preserve these lands for you and others." I nodded silently in agreement, although not seeing the importance of keeping lands yet working so hard to make ends meet whereas he could have sold a piece of land to make things much easier for himself and us.

My father then proceeded to narrate a funny story about my grandfather, his father,

"Once Baba and another man both claimed ownership over the piece of land in Kong-sah. The traditional council had to visit the land to inspect and take evidence for judgement." Now I was even more attentive to hear what grandpa did. He went on.

"Before the agreed day, Baba took a bottle and filled it with water, went to the land in the night, dug the earth and buried the bottle, then took some stones and marked the spot with. On the day that the council went there with both parties to witness, after much talking, Baba claimed he had buried a bottle in the land many years back. Everybody was shocked at the revelation." My father was smiling as he realized he was telling me some rare history. "After much searching and Baba pretending not to know where the bottle was buried, he finally yelled that he had found the spot. Everyone gathered and after removing the hip of stones, they opened the earth and found a bottle underneath. The man who was challenging Baba on the land gave up and that's how that land was secured." I was fascinated with the story of my grandfather whom I never had the opportunity to meet.

Many years have passed, and I still remember the story that my dad told me, land was grandpa's legacy to us and it dawned on me that my father was simply holding on to such a legacy against all odds so as to pass it on to the next generation.

Land meant so much to our ancestors as it was indispensable for farming and grazing. Passing land to the next generation was one of the greatest legacies that our ancestors left, a tradition which goes on till date even as change comes, and more and more people get a livelihood from other activities.

Throughout history, the desire of many leaders has been to leave a great legacy. Alexander the Great's memory is probably more alive today than it was many years ago. Did Alexander know he will pass away some day? Absolutely yes, but his desire to build an empire under his kingdom meant so much to him. He lived to leave a legacy that will stand the test of time.

I sometimes wonder what crossed Mansa Musa's mind as leader of the Timbuktu empire. To pass on to the next generation an empire greater and bigger than what he inherited from his father? Today, many writings celebrate Mansa Musa as the greatest of all amongst the rulers of the Timbuktu empire.

It was King Shaka who made the Zulu people more popular through his conquests in the southern African area, modernising his army which had little rivalry in his day.

Menelik II of the Ethiopian empire is one of Africa's Kings held in high esteem by history texts. Not only did he resist the Italians and enlarge his empire, but he was also recorded as a king who had the people at heart, especially the poor. Appointing administrative advisors who were trusted and respected nobles to help in the administration of the empire earned him much respect and he was widely called "Emiye Menelik" in Ethiopia for his forgiving nature and altruism.

Up to this time, we still remember all those leaders who left an indelible mark on the history of the continent, from those who built the pyramids in Egypt and Sudan to the great empires of west Africa right down to Southern Africa.

When the men and women who stood against colonialism and its ills braced the challenges of the moment and stood up, they knew what they were getting into, yet they did not let fear of going to jail or death deter them. Oppression and exploitation by the colonizers left these people no choice than to inscribe their names in the annals of African history. Their activities eventually resulted in most of them going to prison while some lost their lives in order to leave a legacy. Today, their names are immortalized in the minds of every child of Africa as well as in history books because of their selflessness and bravery. Just like slavery, colonization came to an end because some individuals challenged a status quo they saw as wrong and unacceptable and risked their all for abolition.

Today, Nigerian entrepreneur, Tony Elumelu, is one African creating an impact throughout the continent with the well-known

Tony Elumelu Foundation and other projects. He has understood that helping the underprivileged youths attain their life's goals is an act that will have a ripple effect on the lives of people for generations to come. Through his "Africapitalism initiative' he hopes to make a difference. Mr. Elumelu insists that Africapitalism is not capitalism with an African twist; it is a rallying cry for empowering the private sector to drive Africa's economic and social growth. And some day, long after the gentleman from Jos in Nigeria's Plateau State would have passed, his memory will live even stronger in millions of households because of his great works!

A few years after decolonization was presumably completed on the African continent, the euphoria of self-rule was however short-lived. The Gold Coast (Ghana) led the decolonization train of Sub Saharan African countries, gaining their independence in March of 1957 under the leadership of Nkwame Nkrumah. Ghana's independence sparked a wind of change that would blow from the West African colony to all directions on the continent. Nkrumah stood before tens of thousands of supporters and proclaimed, "Ghana will be free forever."

Speaking at the first session of the Ghana Parliament on Independence Day, he told his new country's citizens that "we have a duty to prove to the world that Africans can conduct their own affairs with efficiency and tolerance and through the exercise of democracy. We must set an example for all Africa." Nkrumah was hailed as the Osagyefo – which means "redeemer" in his native Akan language. The Ghanaian independence became one of the most internationally reported news events of African history according to the BBC world service archives.

The first to obtain her independence, the Independence Day saw foreign dignitaries from the UK, the US, the USSR and hundreds of reporters who'd come in to witness the event. Ghana opted for a red, gold and green flag with a black star at the center. Red stood for the bloodshed of those who shed blood to obtain independence; the gold represented the mineral wealth of the country, while the green symbolized the country's rich forests and natural wealth, and the black star, the symbol of African emancipation.

Ghana's was to be the colors of the flag that most West African nations would copy at independence with alteration of color and star positions. Ghana's flag drew its inspiration from the Ethiopian flag with the red stripe standing for power, faith and blood; yellow symbolizing the church, peace, natural wealth and love; and the green representing the land and hope.

Ghana established herself as a beacon of decolonization throughout Sub Saharan Africa and anti-colonial movements quickly drew their inspiration from Ghana and her president. What happened in the newest nation was definitely going to affect nations born after her. The 1960s saw a lot of new states emerging from colonial rule.

While the new countries followed Ghana's lead, Nkrumah had started feeling threatened and longed to centralize power. In 1962, three younger members of his Convention People's Party "CPP" were reportedly brought up on charges of taking part in a plot to blow up Nkrumah's car in a motorcade. The only evidence against the alleged plotters was that they rode in cars well behind Nkrumah's car. When the defendants were acquitted, Nkrumah sacked the chief judge of the state security court, and then got the CPP-dominated parliament to pass a law allowing a new trial in which all the three men were convicted and sentenced to death, though these sentences were subsequently commuted to life imprisonment.

Shortly afterward, Ghana's constitution was amended to give the president the power to summarily remove judges at all levels. In 1964, Nkrumah proposed a constitutional amendment which would make the CPP as the only legal party in the country, with Nkrumah as president of both nation and party. The amendment passed with 99.91 percent of the vote, a doubtful high total that led observers to condemn the vote as "obviously rigged." The Osagyefo (redeemer) was gradually becoming the oppressor, with a growing thirst for absolute power. Having lived in some of the best democracies in the world, the United States and Britain where he got his higher education, one would have thought that Nkrumah had learnt that rotatory power is better since it allows all who aspire for leadership positions to pursue such and only obtain them by merit and not through force or favoritism. Rather, Nkrumah drew his inspiration from Russia to build a personality cult with an iron fist on power.

Nkrumah's power consolidation was uniting some of his rivals while distancing some of his collaborators as well. By 1966, army officers: Col E.K Kotoka, Major A.A. Afrifa, Lieutenant General J.A. Ankra, and Police Inspector General J.W.K. Harlley justified the military coup which abruptly ended Nkrumah's presidency. They charged that the Nkrumah's CPP administration was abusive and corrupt, pointing to the absence of democratic practices in the nation. The military coup of 1966 was a nationalist one because it liberated the nation from Nkrumah's dictatorship, a declaration that was supported by Alex Quaison Sackey, Nkrumah's former minister of foreign affairs.

Had Nkrumah handed over power through universal suffrage as he came in, his legacy in Africa would have been undiluted. His strong hand on power surely set a precedent and inspired more strong men to hold on to power and subsequent military coups which West Africa will then go on to witness for decades. The first coup might not have been staged in Ghana but her position and role at the time were a catalyst which influenced a lot around her.

It was the mistreatment of indigenous cocoa farmers that urged Houphouet - Boigny to start complaining against the French in Ivory Coast. In 1932, he decided to act, leading a movement of farmers against the influential white landowners and for the economic policies of the colonial government, who favored the farmers. He published an article titled "On nous a trop volés" (They have stolen too much from us), in the Trait d'union, an Ivorian socialist newspaper. It was published under a pseudonym. He would later on lead Ivory Coast to independence with the same well-known promises, which unfortunately never turned out to be respected. He eventually transformed the country into a single party rule, opposition was completely silent as he made himself president for life.

Despite the fact that Houphouet-Boigny left behind infrastructure like the Institute Polytechnique and an international airport, they were nowhere compared to the 320.000 sq ft by 518 ft Basilica of our Lord of peace he built in Yamoussoukro at the cost of $150 - 200m. He wanted to satisfy his desire for his country to have the biggest church in the world, bigger than St. Peter's Basilica in Vatican City, the largest church in the world at the time. Despite the fact that his country lacked healthcare, education and general social amenities, it was the pride of having the biggest church building in the world in his hometown Yamoussoukro that was of priority to him. The basilica was referred to as “"the basilica in the bush." by some Western leaders.

Just as Nkrumah and Boigny, the men who put up a great fight against colonialism turned to tighten their grip on power without a sense of what will transpire when they bow out to nature. Perhaps they cared little about their legacies.

As each strong man emerged from the shackles of colonialism, it dawned on him to seize power then, build a strong wall of state-of-the-art security which catered for him. Some of the few men who emerged from the chains of colonialism with an upright attitude were only to realize later that their country men had betrayed them to be slaughtered in a brutal manner.

Though they used the ills of colonialism and gained the people’s trust, it was rather in the colonial masters’ land that they transferred wealth which was meant for the people. Even today, billions of US dollars are stored in Swiss banks against the names of countless

African strong men who have occupied office. Successive Nigerian presidents have recovered some billions of dollars from Switzerland which the one-time military ruler, Sani Abacha transferred there while the country drowned in poverty. Billions still remain in Western accounts, kept by African leaders and their fellow administrators.

When Togo's first president, Sylvanus Olympio was murdered, allegations were made that his counterpart to the west, Kwame Nkrumah aided the coup plotters. Gnassingbé Eyadéma who led the coup will later on seize power and rule the small west African nation for 37years till his assassination on february 5th, 2005.

When Nkrumah was overthrown, Houphouet - Boigny was alleged to have been instrumental in the coup. He is also believed to have contributed to the assassination of Thomas Sankara. Africans had turned to betray their fellow Africans to the same colonialists that they all fought to gain independence from.

Dictatorship was and is still the order of the African political landscape from Libya to Zimbabwe, from Uganda to Gambia; widespread corruption and poverty increased by the day. The African leaders have found solace in the same people, the Europeans who mistreated their people and have rather caused agony on the indigenous people. In the end, we are our own enslavers first. Only we can save ourselves, living to leave a legacy.

I once listened to the BBC reporting of Switzerland's repatriation of over a billion dollars of looted money that the former military leader, Sani Abacha kept in Swiss banks. The then president, Olusegun

Obasanjo had worked hard to see that money transferred to offshore accounts be sent back to Nigeria. All governments after Obasanjo right up to the present leader Muhammadu Buhari have succeeded in getting back some of Abacha's stolen money. News about stolen money by African strong men was no surprise to our ears as we grew up.

In secondary school, we usually argued in our gatherings to know which one amongst the strong men of Africa is the richest. Even without any proof to back our claims, one of us had proclaimed that "Mobutu of the Democratic Republic of Congo will always be the richest of all", another then countered confidently that Abacha of Nigeria had no rival amongst the strong men. "Abacha didn't rule as long as Omar Bongo of Gabon, that old man is the richest. The population of Gabon is smaller compared to other African nations, they have huge reserves of crude oil, and he controls it as his personal property." Another person gave reasons as to why the powerful man of Gabon is the wealthiest. We explored all of them, from Hosni Mubarak of Egypt to Robert Mugabe of Zimbabwe. Their drive to transfer money into foreign bank accounts seemed to me as if they competed amongst themselves as to who could transfer money out of Africa more than the rest of them.

As the strong men syphoned money from the continent, they equally inculcated the spirit of looting in their subordinates. As the strong man of the land grabs his share, the second in command does the same and as an Egyptian pyramid from top to bottom, the nation's money ends up in the pockets of those who assume positions. In effect, what the leader looted could never amount to the amount that those below him eventually stole.

Officials who fail to get a share of money directly from the state's coffers resort to requesting bribes from the common man on the streets. As such, one of the biggest problems facing Africa today is the custom of enriching oneself which has been instilled in Africans by most of the so-called fathers of the nation. It is often echoed today in Africa that "the easiest way into wealth is by getting involved in politics or church." Both careers, though being "servant careers," have unfortunately been turned into easy money-making careers.

Growing up in poverty with no clear path to making a decent life, with the examples set by the founders of the nations, the coming generations who lack the much-needed honesty have gracefully followed in the footsteps of unscrupulous leaders. They buy their way to the positions they desire as such positions have been put up for sale by the people in charge. In a few months we watch them metamorphose before our very own eyes into pot belly and double chin monsters with an insatiable desire and greed that crumbles their young nations. The poor and those who desire to live an upright life have little choice than to live in misery caused by their greed, corruption and embezzlement, among other ills that erode their legacies before they even go to the land of their ancestors.

CHAPTER 10

THE LAND OF VANISHED DREAMS

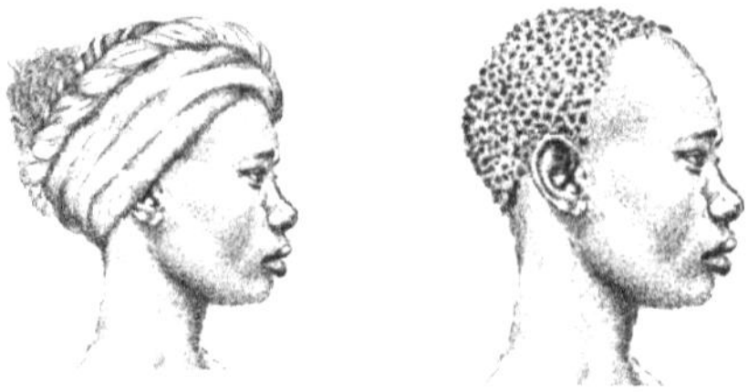

CHAPTER 10

THE LAND OF VANISHED DREAMS

As African leaders have occupied themselves with amassing wealth for themselves and transferring the nations' wealth to Europe and America, they have as well transferred the hopes and aspirations of millions of youths all over the continent with it.

Two of Africa's biggest economies, Nigeria and South Africa along with other Sub Saharan African countries have some of the highest unemployment rates in the world. [1]

Graduating from an institution of higher learning and expecting to pick up a job in a Sub Saharan African country can be as much of impossibility as finding water in the Sahara Desert.

While senior civil servants way above retirement age hold on to positions they should have ceded to the younger generations, younger people leave higher institutions trained to rather compete with the unskilled for menial jobs, these graduates are willing to accept anything that will keep them afloat even if such is completely unlike what they trained for. The few young people who get jobs realize that they are constantly lacking in many aspects and the salary from such jobs cannot sustain them. In most cases, the salaries are insignificant because they simply go to pay off debts accumulated in the past. The abundance of talent and ambition posed against an acute lack of opportunity constitutes a ticking time bomb for Africa and the world.

The rates of unemployment in Sub-Saharan Africa are amongst the highest in the world. This has forced the younger generation to seek greener pastures in other places all over the world, sometimes going through some of the most difficult routes to seek these pastures. These desperate journeys that African youth take are so dangerous, they are putting their very lives at stake. Nevertheless, the promise to those who survive outweighs all the risks involved. The greatest wish of a majority of the African youth today is to make it to the West as life gets harder each passing day.

At the elementary level, most African children have dreams like children everywhere else. Unfortunately, these dreams are dashed as they grow up and face the harsh realities of life in countries where they have been sold out by their leaders. Their dreams of taking burdens off their parents' shoulders and dreams of starting a new life seem to be drifting further and further away, out of sight as they age. It is not surprising that they are resolved to leave, no risk being too high for the youths of Africa to take as leaving the continent through whatever means still remains much better than staying.

Today, I live in the United States, never a day goes by without me thinking of the millions of young agile Africans who long to be where I am so as to have a more deserving life. A day never goes by without me thinking of the uncountable number of my fellow Africans who have perished in the Mediterranean Sea and in the South American jungles in their quests to escape adversity at home and bring some fulfillment to the families that gave everything to educate them. Video and picture footage of fellow Africans dying in jungles in South America are replete on social media. Such images have come to be so familiar that no one is shocked by their horror anymore.

The hardship on the black continent brought about by neocolonialists, with their phony backed governments that have run the continent from independence keeps forcing able Africans to flee.

Immigration has now become a major political issue that keeps coming up in the Western political discussions.

In January of 2019, the Italian deputy prime minister, Luigi di Maio accused France of fueling migration by damaging the economies of the African states. "The EU should impose sanctions on France and all countries like France that impoverished Africa and make these people leave because Africans should be in Africa, not at the bottom of the Mediterranean. "If today we still have people leaving Africa, it is due to several European countries, first of all France, that didn't finish colonizing Africa" Di Maio, Italian deputy Prime minister said to reporters in 2019 while raising a point about the migrant crisis in Europe.[2]

The Italian deputy prime minister's remarks had not come because of sympathy for the Africans. Rather, he spoke thus because the migrants had become a burden on the Italians since their country is situated as one of the first points, receiving those fleeing from Africa into Europe. He also made the comments because the European Union had agreed on shared responsibility to relocate the migrants amongst their various states but some countries like the Czech Republic, Hungary, Poland, and Slovakia refused to support any form of responsibility for migrants arrivals. The inability to relocate migrants amongst themselves has left Italy feeling the weight of these stranded people, and as such they were frustrated with the situation.

Luigi di Maio further remarked that "if it wasn't for Africa, France would rank 15th among world economies, not in the top six." Apparently, the French are the most exploitative of the neo-colonialists, their policies on the continent have stolen the hopes of millions of Africans. African leaders who took stands to challenge the unending imperialist tendencies of the former colonial powers have generally been met with overthrowing their rule or even death. France has never been hesitant to eliminate presidents who posed a threat to their imperialist plans, most recent cases being Marc Ravalomanana of Madagascar and Laurent Gbagbo of Ivory Coast who were deposed in March 26, 2009 and April 11, 2011 respectively and Muammar al-Gaddafi of Libya who was killed in October 20, 2011. France has consistently connived and been in collaboration with other Western powers to assassinate African presidents they deem a threat to their interests. The African

Exponent has compiled a list of presidents allegedly assassinated by France since 1963.

- **IN 1963: SYLVANUS OLYMPIO, PRESIDENT OF THE REP. OF TOGO**
- **IN 1966: JOHN-AGUIYI IRONSI, PRESIDENT OF THE REP. OF NIGERIA**
- **IN 1969: ABDIRACHID-ALI SHERMAKE, PRESIDENT OF THE REP. OF SOMALIA**
- **IN 1972: ABEID-AMANI KARUMÉ, PRESIDENT OF THE REP. FROM ZANZIBAR**
- **IN 1975: RICHARD RATSIMANDRAVA, PRESIDENT OF THE REP. OF MADAGASCAR**
- **IN 1975: FRANÇOIS-NGARTA TOMBALBAYE, PRESIDENT OF THE REP. OF CHAD**
- **IN 1976: MURTALA-RAMAT MOHAMMED, PRESIDENT OF THE REP. FROM NIGERIA**
- **IN 1977: MARIEN NGOUABI, PRESIDENT OF THE REP. OF CONGO-BRAZZAVILLE**
- **IN 1977: TEFERI BANTE, PRESIDENT OF THE REP. ETHIOPIA**
- **IN 1981: ANOUAR EL-SADATE, PRESIDENT OF THE REP. FROM EGYPT**
- **IN 1981: WILLIAM-RICHARD TOLBERT, PRESIDENT OF THE REP. OF LIBERIA**
- **IN 1987: THOMAS SANKARA, PRESIDENT OF THE REP. OF BURKINA-FASO**

- **IN 1989: AHMED ABDALLAH, PRESIDENT OF THE REP. COMOROS**
- **IN 1989: SAMUEL-KANYON DOE, PRESIDENT OF THE REP. OF LIBERIA**
- **IN 1992: MOHAMMED BOUDIAF, PRESIDENT OF THE REP. ALGERIA**
- **IN 1993: MELCHIOR NDADAYÉ, PRESIDENT OF THE REP. FROM BURUNDI**
- **IN 1994: CYPRIEN NTARYAMIRA, PRESIDENT OF THE REP. OF BURUNDI**
- **IN 1994: JUVENAL HABYARIMANA, PRESIDENT OF THE REP. OF RWANDA**
- **IN 1999: IBRAHIM BARRÉ-MAINASSARA, PRESIDENT OF THE REP. OF NIGER**
- **IN 2001: LAURENT-DESIRED KABILA, PRESIDENT OF THE REP. CONGO-KINSHASA**
- **IN 2009 JOÃO BERNARDO VIEIRA, PRESIDENT OF GUINEA-BISSAU**
- **IN 2011: MUAMMAR GADDAFI, PRESIDENT OF THE REPUBLIC OF LIBYA.**[2]

While African leaders who go against France and their allies' interests are eliminated, the proxy leaders on the other hand rule their respective countries for life and with little or no regard for the indigenous people. A majority of the leaders on the continent are masquerading as leaders but their role is to cover up for the activities of the imperialists. Such leaders are given numerous opportunities as a reward of being loyal to the masters, their prime mission being to keep the masters satisfied against all odds. The suffering of the people means nothing to them so long as the masters are happy, and they loot their own share and store in the masters' countries.

As the issue of migration persists due to Africa's foolishness on the one hand and greed of the Western world on the other hand, the big cats as always keep proving their intelligence. The people of the continent which they exploit do not mean as much as the materials they rip from the continent.

The European Union Border Assistance Mission (EUBAM) was launched in 2005. It promotes border control, customs and trade norms and practices that meet EU standards and serve the needs of its two partner countries. It is an advisory, technical body based in Odessa (Ukraine). It has a Liaison Office in Chisinau and six field offices; three on the Moldovan side of the joint border and three on the Ukrainian side.

On 22nd May 2013, the Council of the European Union gave the green light for EUBAM Libya, a civilian Mission under the Common Security and Defence Policy (CSDP), to support the Libyan authorities in improving and developing the security of the country's borders. The budget of the Mission from 22nd August 2016 to 21st August 2017 was €17 million according to reports from the EUBAM and EU websites.[3]

However, due to the political and security situation in Libya after the assassination of Gaddafi in October 20, 2011, EUBAM has been operating from Tunisia where they monitor and stop boats from taking African migrants to Europe.

According to the European council on foreign relations, an approximate of one million Africans made it to either Italian or Greek shores in 2015, and by 2017, the number was curbed to just about two hundred thousand migrants. Collaboration amongst the Western powers has limited migration from the continent of hopelessness. The imperialists have invested more in keeping the Africans in their land while turning a blind eye to the reasons that keep chasing them from home.

With anti-immigrant sentiments on the rise in the Western countries, their politicians have also taken measures to keep the migrants far from them. According to information from the site of the European council on foreign relations, European governments have tried to provide immediate solutions to what is a long-term, structural shift. In this context, Niger became one of the first nations that the European Union engaged with, due to its position at the heart of migration routes to the Mediterranean. European countries' activism in Niger involved the deployment of French and US troops there, along with activities at a German base in Niamey and the deployment of Italian soldiers in Madama (the last outpost for migrants traveling north before they reach Libya).

Interestingly, the Europeans seek solutions to the problems that immigrants pose to them, but they fail to seek solutions to the problems that triggered these very people to leave their homes. With their forces guarding their borders, paying huge sums of money to the northernmost countries of Africa to keep Africans from leaving as well as deploying their military on African soil to stop them from leaving while maintaining their policies of exploiting and supporting the strong men of the continent, Africans if wise enough would reason and work together for their common good.

The Democratic Republic of Congo is widely considered to be the richest country in the world regarding natural resources; its untapped deposits of raw minerals are estimated to be worth in excess of U.S. $24 trillion. (4)- https://globaledge.msu.edu. But, it is also the children of this rich land who undertake risky missions across deserts and jungles and the raging waters to find a better life for themselves and their families while their rich home lies in ruins since imperialism has no conscience for lives that aren't of their race.

It is the children of the former French colonies whose countries continue to pay colonial taxes to the colonial master (France) after more than half a century of independence. While the French continue to use these countries as personal property, with phony presidents, indigenes left with nothing but to search for a better life out of home while the French have never relented in their exploitation of the resources found in these countries with little or no compensation. It is also the children of Mother Africa who continue to bear the weight of imperialism, who have resorted to flee to all directions of the world chased away by hardship brought about by the wickedness of western imperialism, coupled with sheer folly from us.

At the end of a tedious day working under extreme cold temperatures in the West, I couldn't help but think of the West's greed on one hand, and our foolishness on the other. I wondered if at all there will be a day when the powerful buffaloes will come to their senses and realize the strength they have in their muscles and use this strength to demand the big cats to start treating them with respect, then hope assured my imagination that surely the buffaloes will someday, this will happen.

However, fear also engulfed me that perhaps the maker of everything made the buffaloes and all like them to be used by the big cats for the rest of their lives as this is his plan for the ecosystem whereby, some though endowed with great riches and number, are not endowed with intelligence and are bound to be used by the more intelligent for the rest of their lives. And if we are bound to be this foolish forever, then, the misery will just go on and on, the Mediterranean Sea will keep on consuming the offspring of the 'Black Continent' in their attempt to make it to Europe while the jungles of South America keep nourishing their roots with the remains of those who fall en route to North America in their pursuit of a better life.

CHAPTER 11

THE UNLEARNED LESSONS

CHAPTER 11

THE UNLEARNED LESSONS

In the jungle, it is indisputable that buffaloes, giraffes, zebras and all the big animals that fall prey to the big cats and other organized scavengers like wild dogs and hyenas, do so partly out of their lack of wisdom to realize the strength in their numbers. Though few in number, these carnivores possess the brainpower that makes them superior over their fellow grass eating animals that lack the mental ability to unite though more physically powerful. Power, therefore, lies in the hands of those who reason alike, compromise, have a common goal and uphold each other's interest. Even when such disagree amongst themselves on certain aspects, they always unite when it concerns their common interest or against a common enemy.

It was the common interest in acquiring overseas territories that made the interested European powers assemble in Berlin, in 1884 to discuss the partition and acquisition of territories in Africa. With the outdated slave trade, it was time for them to resort to other means-keep the Africans at home but still use them to exploit the resources of their land, for themselves. The assembly of 1884 didn't mean the European powers had no conflict amongst them, they certainly had tensions, but in order to take over the African race and land, they had to come together to achieve the shared interest.

The age of rapid industrialization ushered in the growing powers of Britain, France and Prussia/Germany, and, later on, the United States and Japan. This led to imperialist and colonialist competitions for influence and power throughout the world, most famously the scramble for Africa in the 1880s and 1890s. Its reverberations are still widespread and consequential on the continent in the 21st century.

It was also during this epoch that alliances began forming within Europe: The Triple Entente formed by France, Britain and Russia and the Triple Alliance formed by Germany, Italy and Austria-Hungary.

It is this lesson of working in unison when tackling a common problem that many Africans have failed to learn. After losing a great number of ours to the evil deeds of slavery, suffering under the degrading realities of colonialism and continuous neo-colonialism, we still cannot confront the most challenging problem that Africa has been facing for centuries. Post independent Africa has failed greatly due to the mistake of taking on a unified Western world individually rather than in common. Most of the pioneer leaders concentrated more on their various countries, concentrating powers to themselves while leaving space for the imperialists to take over the continent. Dr. Kwame Nkrumah stated;

"Our independence is meaningless unless it is linked up with the total liberation of Africa." "Seeing you in this... it doesn't matter how far my eyes go, I can see that you are here in your millions and my last warning to you is that you are to stand firm behind us so that we can prove to the world that when the African is given a chance, he can show the world that he is somebody!"

"We have awakened. We will not sleep anymore. Today, from now on, there is a new African in the world!"[11]

These excerpts of Kwame Nkrumah's speech at independence, words which would have made much sense if he and his contemporaries had reasoned alike and worked together towards building a new Africa, cutting off from European domination.

Rather, Nkrumah collided with Sylvanus Olympio on a number of issues and the two could not work towards their common goal or against their common foe. Just like his closest neighbor to the East, Sylvanus Olympio, to the West Nkrumah still could not find a friend to work with in Félix Houphouët-Boigny of Ivory Coast. While Nkrumah was accused of aiding the assassination of Sylvanus Olympio of Togo in 1963, Boigney was suspected of aiding the overthrow of Nkrumah in 1966 and the assassination of Thomas Sankara of Burkina Faso in 1987.

According to multiple writings, Houphouet-Boigny, besides supporting policies pursued by France, also influenced their actions in Africa. At his death, the Radio France International (RFI) described him as the "father of France-Afrique"

The Washington Post's editorial of December 8th, 1993, after his death stated that he supported the Biafran war of independence, a war which France was said to be in favor of in order to have access into the oil reserves in the Biafra land. Houphouet-Boigny pushed France to support and provide arms to warlord Charles Taylor's rebels during the First Liberian civil war in hopes of receiving some of the country's assets and resources after the war. Any leader who stood against France in Africa automatically was at odds with the Ivorian leader who leaned strongly towards his colonial master.

The struggle for independence had been a burdensome one, the colonial masters unwilling to let the indigenous people of Africa take charge of their land and themselves. The unwavering will of the people behind their leaders to take over homeland, led to the acceptance to hand over to the people. After much bloodshed and death and then the imminent end to colonial rule, the African leaders then failed to come together. Their individual strength then became futile against an organized West. Even more saddening, was the fact that some African leaders instead teamed up with the enemies of Africa to eliminate their fellow African brothers.

Of all assassinations of African leaders, from Togo's Olympio in 1963 to Libya's Muammar Gaddafi in 2011, it has been the involvement of a few Africans with the Western powers. Even a full grown elephant which fails to unite with its own will fall prey to a united pack of wild dogs. While the West comes in as one against

Africa, African nations face the united Western front individually and the results for over half a century have been nothing short of disgraceful.

"Each time I want to fight for African rights, I use only one hand because the other hand is busy trying to keep away Africans who are fighting me." Benjamin Burombo's words explained the whole reason why the African puzzle cannot be fixed.

Julius Nyerere envisioned an East African community of Tanzania, Kenya and Uganda. He even opted for Tangayika's independence to be delayed till the dates of Kenya and Uganda, a suggestion opposed by his very own party and later rejected by the new presidents of independent Kenya and Uganda. Nyerere endorsed the Pan-Africanist idea of unifying Africa as a single state, although he disagreed with the Ghanaian President Kwame Nkrumah's view that this could be achieved quickly. Instead, Nyerere stressed the idea of forming regional confederations as short-term steps towards the eventual unification of the continent.

The newly independent states on the 'Dark Continent' however came up with the Organization of African Unity (OAU). They were thirty-two African states, with the main aim of bringing the African nations together to resolve the issues within the continent. They met in Addis Ababa, Ethiopia on 1 May 1963. Alieu Ebrima Cham Joof, the late Gambian historian and Pan-Africanist delivered a momentous speech which however fell on deaf ears in Ethiopia.

"It is barely 75 years since the European Powers sat round the table in Germany each holding a dagger to carve up Africa for its own benefit.... Your success will inspire and speed up the freedom and total independence of the African continent and eradicate imperialism and colonialism from the continent and eventually neo-colonialism from the globe... Your failure, which no true African in Africa is praying for, will prolong our struggle with bitterness and disappointment. I therefore adjure that you ignore any suggestion outside Africa and holding that the present civilization, which some of the big powers are boasting of, sprang up from Africa, and realising that the entire world has something earthly to learn from Africa, you would endeavor your utmost to come to agreement, save Africa from the clutches of neo-colonialism and resurrect African dignity, manhood and national stability." Alieu Ebrima Cham Joof.[2]

The leaders of the independent African nations left the conference of 1963 with little or no desire to advance the general interest of the African people. The warning to keep aside differences and form a united front to end neo-colonialism was not heeded, the results which followed from Togo in 1963 were rather the sheer opposite. Accusations and counter accusations against neighboring countries trying to assassinate each other's president, the case of Ghana and Togo; Ivory Coast and Guinea were common. As the leaders rather concentrated on silencing opposition and consolidating power, dissidents found refuge in countries where the opposing leaders were at odds.

The leaders at the time, just like the powerful but foolish buffaloes failed to learn the lessons in dealing with predators. Meanwhile, the predators continue with their intelligence and have progressively developed new and smarter tricks that keep luring the thoughtless buffaloes to keep falling prey while they are being fed on. And though the buffalo's backyard is so rich that it could have let her and her type to live as kings and queens with their princes and princesses, they have rather lived a degrading sorrowful life of servitude forever.

When the Duikers of Zimbabwe decided to muscle up with the lions that came in from England and seized their land, they had an advantage in numbers against the invading lions. After putting on a strong resistance through guerilla warfare against the colonial lions, they succeeded in toppling the lions' minority government. Next thing was to take back their lands which the lions had occupied by force. This policy of land reclamation which was not accepted by the lions who controlled the Zimbabwean farms drew support from the world of the big cats, with their intelligence of a unified force they sanctioned Zimbabwe in every possible way while African countries stood aside and watched Zimbabwe suffocate to near death.

I admire how these predators unite even amidst their differences when it concerns a common enemy. While the whole predator world united against Zimbabwe with their sanctions, the fellow cracked-brained prey rather couldn't support Zimbabwe in any way. Apart from South Africa that stood with Zimbabwe, probably because of their shared border or common history of resistance, some African preys rather collaborated with the predators against their type.

History tells us that no people in captivity or oppression ever got their freedom by trusting in the goodness of the oppressors, even our people who stood up against colonialism did not wait for the colonizers to get fed up with colonial rule and then hand control to the indigenous people. It was the indigenous people who under the cruelty of colonialism meted on them by the Europeans, rose up and demanded its end. And even after demanding its abolition, the white world came up with neo-colonialism and imperialism which still continue to suffocate Africa today.

Thomas Sankara, unhappy and uncompromising with the conditions in which his country and oppressed people throughout the world found themselves in, challenged the oppressed people to stand up for themselves. He famously said, “The slave alone will be responsible for his own misfortune if he harbors illusions in the dubious generosity of a master pretending to set him free. Freedom can be won only through struggles, and we call on all our sisters of all races to go on the offensive to conquer their rights.”[3]

Sankara was fed up with promises made by the West, promises to uplift the African continent and reduce misery in her people. Regrettably, the majority of the African people had not learnt the trick of coming together against the ills of imperialism, and for over fifty years of continuous exploitation by the West and now China, the continent with abundance cannot find the wit to come together for the betterment of their people.

Prof PLO Lumumba believes Africa must rise from slumber in order to thrive. Says he, “Africa must define her own problems, and after defining her own problems, must also find solutions to her own problems. Do not be cheated, do not be diluted that the world is full of benevolent individuals and institutions whose only claim to fame is that they want to lift Africa from the morals that she now finds herself in. The reality is the poverty of Africa is the glory of some civilizations. Poverty is an industry, ignorance is an industry, big industries and there are those who want Africa to continue to wallow in poverty and ignorance because it is their safest bet to their continued benefit. The day we wake up to that reality, the safer we are. That is why the English have this beautiful saying, "let sleeping dogs lie" so we're sleeping, and we're being allowed to lie, and we're being hypnotized into even greater sleep and slumber"[4]

One thing Africans haven't learnt according to Lumumba is keeping the Western world out of their issues and seeking solutions by themselves. Lumumba's words could be wrong if recovery measures implemented by bodies like the world bank and the IMF had improved the quality of life on the continent if loans and aid from the international community had had positive outcomes on the continent. Such measures have rather had no positive bearing on the continent as Africa continues to drown in debts and relies on aid. "There is no free meal", Lumumba laments about aid that Africa blindly depends on year after year.

Today, I see numerous activists from across Africa who challenge imperialism and neocolonialism just as the leaders of yesterday who challenged colonialism did. The likes of PLO Lumumba, Arikana Chihombori-Quao, Nathalie Yamb, Kemi Seba, Julius Malema, Chumani Maxwele all speak passionately about Africa's future divorced from her exploitative past. The reasons that have provoked these children of Africa to speak up aren't different from the grounds that anti-colonial activists stood up and subsequently brought an end to colonial rule. Each time I listen to these fine children of Africa speak, my expectations get renewed that all isn't lost for the continent.

Unfortunately, just as the first activists failed to unite in their common objectives and later struggled independently against imperialism, today's activists seem to be the same. With the help of technological advancement in this era, we easily follow them as they deliver impressive talks pertaining to Africa's problems. Never, have I ever watched these admirable children of Africa come together to brainstorm on how to mobilize African people to challenge the system that continues to enslave their people.

Hundreds of years ago when Williams Wilberforce took up the quest to challenge slavery, he realized that just talking to the politicians of his time would not bring effective impact. He got the British people involved, sugar which was produced by slave labor was boycotted by people of good conscience. He lobbied for petitions to be signed and sent to parliament which was successful. When businesses witnessed drop in sales, politicians who continued to support slavery risked losing their seats due to Wilberforce's campaigns, they realized that change was imminent.

In the year 2020, after so many years of crying injustice, constant police brutality and killings, the African American population in the United States gained worldwide sympathy as George Floyd begged for dear life while forced on the ground with a knee on his neck. The world watched in shock as a man was killed by a law enforcement officer. People of godly conscience of all races and creeds, across borders, denounced the harsh realities that African Americans go through in the much acclaimed 'Free World.'

Politicians who had turned a deaf ear to the cries of the African American population concerning police brutality and other forms of injustices are today coming to terms as Americans of all races are gradually sympathizing with the African American minority. The campaign "Go out and vote" has seen politicians who refused to accept the "black problem" voted out, sending a strong statement that the "Black lives matter" and if not tackled, such politicians would have no place in today's political landscape of the United States.

Should the activists who often denounce French policies in Africa, who denounce neocolonialism and imperialism come together as a fist rather than fingers and galvanize the people in movements and protests, then the cries of the Africans will fall on the ears of some people of good conscience.

We have to learn the importance of unifying and when we do, then the millions of Africans who've passed through thick and thin to get to the West will assemble in major cities of the world to denounce the second scramble for Africa.

The fate of millions of Africans, who dream of escaping the hardship at home depends to a great extent on the African diaspora. Unfortunately, an escape from the dark continent, the 'shit-hole' continent sometimes means forgetting the past (Africa) and concentrating on the present (West). Happenings on the ground no longer worry those who have left, while the "left behinds" continue to suffer the tribulations set on them by the imperial world and hoping for a day that they too will flee from home.

For so long Africans have nursed the belief that the Western world would help them overcome economic and political challenges. We have wished for successive Western governments to make things better on the continent.

In 2008, Africans saw the one they thought was one of theirs, a black man of African descent occupying the White House in the person of Barack Husein Obama. Three years later, this same one in whom most of Africa saw themselves in, collaborated with France's Nicholas Sarkozy to eliminate Muammar Gaddafi of Libya, one of the few African leaders who had often criticised and stood up against Western imperialism. As Libya lays in ruins following this invasion, it further dawns on Africa that the West cannot fix the problems of the continent.

"If western interventions in Africa were to solve problems, the war in the Congo should have long ended" my father always told me when we talked of depending on the Whiteman to solve our crisis. Many analysts of African affairs have voiced out that a majority of the wars in Africa are proxy wars instigated by the Western world to create an atmosphere for them to come in and do the dirty work which continues to hold Africa back, yet Africans are unable to comprehend the same old games.

In the same way that the Italian Deputy Prime Minister, Luigi Di Maio accused France of impoverishing Africa, celebrity entertainers like Trevor Noah, Boris Kodjoe, Akon; writers such as Chimamanda Adichie amongst others have publicly voiced their dissatisfaction at the exploitative policies that Western nations especially France have on the continent.

Kodjoe and Akon are few prominent descendants of Africa born out of the continent who strongly identify with Africa. Kodjoe was born in Vienna while Akon was born in the United States, the former sides with Ghana as his father though being a German, hailed from the Ga-Adangbe people of Ghana. Akon on the other hand sides with Senegal, the birthplace of his parents. While most successful Africans stay indifferent to the imperialist attitudes and policies that make Africa the continent to flee from, these descendants of hers have not stayed mute.

Kodjoe speaks against the fact that former French colonies in Africa continue to pay to France what is called 'colonial debt'. "The debt is $500 billion a year" Kodjoe spoke on 'Here and Now', a public radio magazine program produced by NPR and WBUR.[5]

With public figures like Kodjoe, Akon and a lot more, it is a tool which African activists could use to send out their message to the imperialists as well as to the African diaspora and Africans on the motherland.

According to world meter, Africa is the second largest and second most populous continent after Asia with about 11.7 million square miles including adjacent islands. It covers 6% of Earth's total surface area and 20% of its land area. With 1.3 billion people as of 2018, it accounts for about 17% of the world's human population and Africa's average population is the youngest amongst all the continents.

Globally, Africans in the diaspora are spread out across the continents. In North America, there are 39 million from the African Diaspora; 113 million in Latin America; 13.6 million in the Caribbean; and 3.5 million in Europe, according to World Bank estimates. With the endless hardships on the continent, the number of African diaspora is on a steady rise every year.

While Africans flee home, the predators have not slackened in preying on the African continent. A majority of the Africans in the Western world arrive in the faraway lands without the necessary qualifications to obtain the most prestigious jobs. With so much responsibility waiting for anyone who makes it to the West, furthering education at times becomes a daunting task. This has made lots of these children of Africa continue to live lives of hardship out of home, caring for elderly white folk in senior citizens homes where they sometimes face insults because of their race yet they keep holding on since their current situations appear much better than home.

From the very first day I arrived in the United States, I was curious to know if life is better in the states than at home from the Ethiopian taxi driver who drove me to the hotel. His response wasn't one I'd expected as he explained the stress he and his fellow countrymen in the taxi sector go through. The same question I have asked every

African with whom my paths crossed in the United States and the answers have always been the same, "...if home were a little better than it is, then I won't be here." It is a phrase I have come to be used to from Africans struggling to live the American dream.

"If home were to be better, who would make it better?" I ask myself. Certainly not the French or the Americans or the Chinese as we have known all these years. Whereas successive American governments have made friends with African governments since independence, showing solidarity and love for the continent, Americans of African descent have suffered untold suffering in America and even death. The same has been true in Europe over the years as African people have often been targets of racism with the governments doing little to end such practices but are very active in Africa under the pretense of helping Africa.

During the Vietnam war in the 1960s, the Civil rights leader, Rev. Dr Martin Luther King Jr criticised the sacrifice put up by black people to free people far from home when they themselves had no freedoms at home, America. It is an error by Africans to nurse the hopes that America will help Africa when successive governments, Republican and Democrat alike have failed to help their largest minority population in America ever since slavery was abolished on January 31, 1865. The presence of America in Africa, the aid poured in is just a trap that has enticed and keeps enticing Africans to believe that America has come to help, meanwhile they too are present on the continent for the same reasons as the Europeans who met at Berlin in 1884 for- Africa's resources.

Another thing I have learned among the few Africans that I have interacted with in America is the desire to end the exploitation of Africa of her resources by the imperialists. As I follow discussions on social media bloggers like the New York based Nigerian, Adeola Fayehun among numerous others, there's almost a unanimous conclusion that the West's policies on Africa be changed and corrupt leaders on the 'Black Continent' be brought to justice.

In the words of Thomas Sankara "The Upright man" who was assassinated by western conspiracy in the then "Land of Incorruptible People" "the slave who is not able to take charge of his own revolt deserves no pity for his lot. The slave alone will be responsible for his own misfortune if he harbors illusions in the dubious generosity of a master pretending to set him free." His speech at the United Nations General Assembly on October 4th, 1984. [6]

At this point in history, with thousands of educated sons and daughters of Africa in all fields, with sons and daughters of Africa breaking barriers in science, athletics, the humanities and all that the human being can possibly think of, with millions of us all around the world, in no way can our misfortunes be blamed on other people than us.

In my many discussions with fellow Africans, I have maintained my belief that if the status quo on the continent has to change then it must be the African diaspora to champion such a cause since most Africans live under Western backed dictators who rule the continent with an iron fist and have no regard for African life.

When I watched videos of the protest of August 28, 1963 in which the Rev. MLK Jr. gave the famous "I have a dream" speech, and those of the million man march of October 16, 1995 called by minister Louis Farrakhan, I envisaged the African diaspora in "a million man march" on the streets of major cities of the world demanding an end to the atrocities caused by the western governments on Africans.

CHAPTER 12

THE VITAL ROLE OF "REPATRIATES"

CHAPTER 12

THE VITAL ROLE OF "REPATRIATES"

The International Organization for Migration (IOM) and the Migration Policy Institute have both defined Diasporas as: "Emigrants and their descendants who live outside the country of their birth or ancestry, either on a temporary or permanent basis, yet still maintain effective and material ties to their countries of origin." According to this definition, The Slave Trade was responsible for most of Africa's migration from the 1500s to the 1800s which saw millions of Africans shipped to the Americas and the Middle East.

Globally, Africans in the diaspora are spread out across all five continents.

The U.S. Black population is also growing as other races. In 2019, 46.8 million people in the U.S. identified their race as Black, either alone or as part of a multiracial or ethnic background. That is up from 36.2 million in 2000. The Black share of the U.S. population is higher today than in 2000 as well. About 14% of the national population said they were Black in 2019, up from 13% who did so in 2000.[1]

Unlike the slave trade which took Africans away from home, misery from corruption and imperialism are responsible for African expatriates living mostly in the West and other parts of the world today.

According to The Pew Research, Roughly 4.6 million, or one-in-ten, Black people in the U.S. were born in a different country as of 2019, up from 3% in 1980.

The number of African-born Americans increased from 1,183,316 in 2000 to over 1,676,413 in 2010 according to The World Atlas.

The number of immigrants and their children in Europe represent a potential 2.40 percent of which France has the highest number of the Afro population with eight percent. [2]

Some prominent African descendants have over the years called on fellow diasporas to unite for their common interest in order to fight for the black race which has witnessed untold suffering through various forms of racism and abuse. Amongst them have been the likes of W.E.B Du Bois who amongst other prominent African descendants of his era from around the world assembled in London in 1921 at the second Pan African congress. Du Bios went on to condemn European rule over Africa and advocated for an Africa ruled by Africans and implored the then League Of Nations (LON) to see that Africans be appointed into prominent positions, a request which the LON took very little action on. He disagreed with fellow Pan Africanist, Marcus Garvey who suggested that African Americans return to Africa and rule Africa. Du Bois did not only encourage the African diaspora to return home or visit, but he also spent his last days in the first independent Sub Saharan African nation. Du Bois died in 1963 in Accra, Ghana.

Another important African descendant of the 1920's is Marcus Mosiah Garvey who was born in Jamaica, an orator for Black Nationalism and Pan-Africanism movements for which he founded the Universal Negro Improvement Association and African Communities League. Garvey advanced a Pan-African philosophy which was at times referred to as Garveyism. Garvey's fellow countryman, Bob Marley, through his music encouraged the African diaspora and Africans to unite.

The end of the days of slavery, followed by segregation in the United States encouraged some former slaves to return to their motherland in what was known as the "Back to Africa Movement' in the 19th century. Though it was a failure back in the day, the call for back to Africa is echoing even louder these days, with more and more Afro-diasporans returning to their motherland or visiting. Some celebrated African Americans have returned to the motherland, identified their tribes and even picked up tribal names though not resettling.

Golden Globe nominated actor, producer, director and six-time winner of the NAACP Awards Blair Underwood traced his ancestry to the Vengo (Babungo) people of Cameroon in West Africa. He journeyed with his father to his ancestral homeland in Cameroon and reconnected with his family. In an emotionally laden voice, Blair said at the end in an episode of the NBC show 'Who do you think you are? 'What I will share with my children is that we come from a long line of people from these valleys, to actually be able to travel all these long miles and to meet my cousins face to face, it feels like the family has come full circle to reunite a family that was broken. Voids were filled that I didn't even realize that I had. For me, we are Africans not because we were born African but because Africa was born in us. Who I thought I was when we started this journey is different from who I know I am today. It has been incredible" Blair Underwood spoke to NBC. [3]

In Washington, D.C., in September 2018, Ghana's President Nana Akufo-Addo declared and formally launched the "Year of Return, Ghana 2019" for Africans in the Diaspora, giving fresh impetus to the quest to unite Africans on the continent with their brothers and sisters in the diaspora. This initiative has influenced lots of Afro-Americans from different walks of life to connect with the motherland.

I was listening to the Steve Harvey morning show on my way to work one day, as I always did when Mr. Harvey explained his experience when he visited the motherland. "When I sat in those dungeons where slaves used to be held in man, 'the room of no return' which was the last room that slaves were held in before being shipped, I wept so hard only to realise later on that my son in his corner was there weeping too. It was a great feeling for me being there, it was like my spirits connected with our ancestors in those dungeons, man." said Steve Harvey. "I think every African American needs to visit the motherland," Steve Harvey said on air in a soft and convincing tone.

In 2015, I met an African American lady in Atlanta Georgia who told me of her younger brother who, with his family, had returned to Ghana and started up a business. Showing pictures of his brother and family in Ghana seemed so exciting to her, and she declared how anxious she was to visit home too.

Another native of Atlanta who, like many Afro-diasporans, traced their roots to the continent, is Grammy award winning rapper, Ludacris (Christopher Brian Bridges). With his wife coming from Gabon, Ludacris declared on his Instagram post that he and his entire family were taking up Gabonese citizenship. "Starting my new year off with dual citizenship. Africa, I'm official! Momma and kids too. The best gift of the decade award goes to @eudoxie," Luda wrote. "I am a loyal citizen of Zamunda," he joked. "This is the best day of my life and Wakanda,"

Ludacris in 2019 visited Ghana for the year of the return as posted on his Instagram page a picture of him in the male dungeon. "Our ancestors never gave up faith. You can never imprison our minds nor our spirits. They empowered me to return. I have completed the cycle and I'm beginning a new cycle. The chains have been broken and the indomitable resilience of African people triumphs #yearofthereturn," he wrote in the Instagram post.

When I searched for the rapper on google to find out more about him, it was how the search engine described him that put a smile on my face. “Christopher Brian Bridges, known professionally as Ludacris, is an American-Gabonaise rapper and actor.”

It’s a great thing for Africa. I thought that her descendants are taking pride in her. Afro-diasporans from all over have made trips to a number or of all those countries that Africans departed from.

As African descendants, especially Sub - Saharan Africans continue to be faced with racism in the West, calls to return to the motherland have been on the rise. African athletes in Europe as well as the rest of the Afro-diasporans in Europe especially and in the rest of the world have also endured racism over the years. As these reactions continue to pop up around the world, so too has the consciousness of looking homewards increased.

At a train station in Camden, New Jersey, I met a fellow young African American man. Travis was a worker with the New Jersey transit. I asked directions on how to go about where I was heading. My accent was my identity as it is for Africans in America. After inquiring which part of Africa I come from, he wondered why we would even like to come to America where we will be treated with scorn. "I'm working so hard to save money and return to “motherland” man, America isn't a place for us. I want to be in a place where people won't look and see me differently. Where I'll walk down the street and not be afraid of my skin color being a threat to nobody." Travis told me, with every part of his face showing how serious he was.

Travis’ wishes of returning to Africa have been realised by some African Americans, who faced with prejudices, have made Africa their permanent home. The young Ghanaian youtuber, Wode Maya who goes around Africa reporting African success stories and Afro diasporans who have moved to Africa on a permanent or temporal basis has brought to light a great number of African descendants who have moved to the land which they believe it is theirs, where color is not a factor of their daily challenges.

A number of other Afro-Americans I have come across have shared the same view, as a young lady I met in 2015 back home before leaving for the US that same year. Her parents came to the states from Nigeria. After serving in Cameroon with the United States Peace Corp, she told me she saw a different Africa from what they had learnt in school and a contrast from what the television had told them of Africa. "I'll return to Cameroon as soon as my service with the Peace Corp is over, my few years in Cameroon have been the best of my life and I am at peace here" she said.

During the peak of the lockdown as a result of the Coronavirus pandemic, Africans in some cities in China were a point of racist attacks from Chinese authorities and their citizens. Social media went crazy when video footage showed Africans denied access into shops while some were evicted from their homes, forcing them to be living on Chinese streets and under bridges after being accused of importing the coronavirus. Africans in such cities sent out SOS calls to the world as they were chased from homes they had paid for. Some governments responded by lifting their citizens from China. The actions that took place in China are just some of the humiliations that African descendants have endured around the world for generations because of skin tone.

It is because of the contempt with which the rest of the world treats Africans that repatriates' echoes continue to sound louder and louder.

In Harlem, New York City, a movement began in 1918 known as the Harlem Renaissance whose participants, of African American culture, particularly in the creative arts, and the most influential movement in African American literary history. Embracing literary, musical, theatrical, and visual arts, it sought to reconceptualize "the Negro" apart from the white stereotypes that had influenced black peoples' relationship to their heritage and to each other. They also sought to break free of Victorian moral values and bourgeois shame about aspects of their lives that might, as seen by whites, reinforce racist beliefs.

From the Harlem renaissance in New York City with its cultural awakening, a group of about three men from the former French colonies of Africa and Caribbean writers living in Paris, as a protest against French colonial rule and the policy of assimilation were influenced to begin a literary movement of the 1930s, 40s, and 50s called "Negritude." Its leading figures were Leopold Sedar Senghor, who later on became first elected president of Senegal in 1960, along with Aime Cesaire from Martinique and Leon Damas from French Guiana, who began to examine Western values critically and to reassess African culture.

"The negro has reactions that are more lived, in the sense that they are more direct and concrete expressions of the sensation and of the stimulus, and so of the object itself with all its original qualities and power. This was a reaction against too strong dominance of French culture in the colonies, and against the perception that Africa did not have culture developed enough to stand against that of Europe"

Sedar Senghor wrote on the perception by the French, a belief not limited to the French alone but to a majority of the white race as a whole in view of African culture compared to theirs. Though African consciousness has been gaining grounds ever since, her culture continues to be under threat as Africa still lives under Western hegemony, not able to decide who their leaders will be, not able to take control of their economies and even themselves.

Lilian Thuram, one of the former French colony's citizens from Guadeloupe and member of the team that won the very first soccer world cup for France in 1998, questions the morals of the French with the Franc in their former colonies. "Would it be normal for the Euro to be made in China? I'm sure many French people would disagree. Would it be normal for a certain amount of Francs to be deposited in a bank in China? And whether China decides what the country can spend or not? This is a necessary question.

With this question of the franc of the Financial Community in Africa, Lilian Thuram joins the ever-growing list of public figures who more or less explicitly support the wave of popular challenge for real monetary sovereignty for Africa.

From "repatriates", as opposed to "expatriates", the "repatriates" movement is gaining momentum among the younger generations of the Afro-diaspora. According to the French Development Agency, some 70% of African MBA students in France want to return to

Africa. More and more people know the truths of the African political landscape both past and present. Assassinations and proxy wars that have ravaged the continent for years have been the trick of the West to put a knee on Africa's neck, so that Africa continues to be desperate and keep relying on them. With an increased exposure of what people like Thomas Sankara, Patrice Lumumba, Muammar Gaddafi or even Lorent Gbagbo amongst the more than twenty African statesmen who have been assassinated or dethroned by Western conspiracies stood for, no one today doubts the continuous presence of the Western powers on African soil. The question remains "how will we get them off our necks?"

It is in the backdrop of this French presence in Africa and the consciousness of the activities that repatriates like Kemi Seba, Nathalie Yamb, Akon amongst many have returned to Africa, advocating against the French presence.

Kemi Seba was born Stellio Gilles Robert Capo Chichi in Strasburg, France in 1981 to immigrant parents from Benin. At about eighteen years old, he joined the US based nation of Islam and later formulated his own ideology while visiting Egypt in his twenties. As a result of this process, he abandoned his names at birth and took up the "nom de guerre," derived from Egyptian meaning "black star"

Seba has been at constant odds with the French and their policies in Africa as well as with their stooges as presidents over a number of African countries, his activities have led him locked up for a minimum of four times in France following his never-ending campaigns against the French presence on the African continent.

After returning to Africa in 2011, his campaigns put him at odds with countries like the Ivory Coast, Guinea Conakry, and most recently Burkina Faso where he organized the Burkinabe people in late 2021 to protest the French army's convoy using their country as a gateway to Niger. Though the Burkina Faso authorities barred Seba from entering the small West African country, his mission was however a success as the Burkinabe people even went as far as demanding that their president, Roch Marc Christian Kabore step down for taking sides with the French.

Just as Seba, Nathalie Yamb was born in Gränichen Switzerland but had returned to Africa, precisely to Ivory Coast where she was greatly involved in politics. During a Russia - African summit in 2019, her criticism of French policies in Africa, as well as the government of Ivory Coast didn't go well as she was deported. However, her anti French imperialism activities have continued, especially on social media.

The Western powers, like the predators of the wild, always sense any move that the prey makes and immediately take action to stymy it. The discontent and anxiety of the Afro-diaspora in France has led the French president, Emmanuel Macron to establish the Conseil Présidentiel pour l'Afrique, CPA (Presidential Council for Africa) to revitalize the France-Africa partnership. A large portion of council members are from the diaspora.

In France, diasporans are grouped together in the Forum of Migrants' International Solidarity Organizations (FORIM). Established in 2002, FORIM reflects the determination of migrant associations to better organize themselves under the umbrella of a representative body and the desire of public authorities to have a legitimate interlocutor representing migrants. As a result, the partnership between the diasporans, represented by FORIM, and the relevant public authorities is a recurrent one, with AFD in particular. While Afro-diasporans in France are enticed by the French authorities, the exploitation back home has not abated by one bit.

It is therefore very important for the African diaspora to stand up as one, not as Mali or Senegal, not as Ghana or Togo, not as Nigeria or Ivory Coast, not as Libya or Algeria, not as Rwanda or D.R Congo, not as Liberia or Gambia, not as any one country but as Africa and demand the end to policies that keep exploiting and impoverishing the people. When we, the Afro-diasporans, would converge in major Western cities in our millions and tell of the acts of the French in particular and those of their allies as a whole, we will be on the right side of justice.

Africa's diaspora holds the key of locking up all exploitative policies in Africa and opening up new paths for the people of Africa. Just like the powerful buffaloes, Africans have a secret weapon yet unaware of.

Every single politician in the United States or France or Brazil relies also on the vote of that child from the Dark Continent to propel him or her to wherever they want to be. These millions of Africans will determine who will be president or congress man or woman, they will determine who to occupy any electoral seat at any given level. This is the power that sons and daughters of Africa have held for ages, but they behave as if they have no right to decide who to become an elected official.

Another reason Africans cannot be seen as a force to reckon with is because they do not have a medium through which they can be reached, or their voices be channeled. For as long as I have followed American politics, candidates concerned have always talked about policies towards minorities.

Telemundo serves the Hispanic population in the USA, with their force, concerned politicians give their plan of action regarding Hispanics through this channel. Africans however fall under African Americans, not that they shouldn't fall under the same medium, but because there are different policies that will affect the same ancestral people. If Africans had had a medium which is purely African related, Africans should have voiced out what they expect of each candidate in relation to policies affecting Africa when he or she desires office.

Africa's greatest strength to end this inhumane treatment of her people is the diaspora's number to stand up and challenge the consciences of the good people in the Western world. On his quest to end slavery, William Wilberforce, we are told, exposed how some people made money from black people's blood in the plantations. He exposed the conditions in the sugarcane plantations and challenged the British people to rather consume sugar produced from free labor as opposed to that produced from slave labor. His continuous campaigns gained grounds gradually as more and more people boycotted sugar produced from slave labor.

In the same light, a change in policy by the imperialists will only come if Africans expose the deeds of these governments, how they plunder their resources through maintaining dictators in power while assassinating those who have a clear vision for Africa.

Just as millions of Western citizens were unaware of slave conditions in plantations in the days of slavery, some Westerners have never comprehended why African countries continue to wallow in abject poverty because politicians who put such policies in place never disclose them.

No one could have heard or had interest in the tiny West African country of Equatorial Guinea had they not discovered offshore oil some years ago. But as it goes in Africa, blessings of natural resources have been a gift of the ruling elites while the rest of the people languish in abject poverty. The Western powers come in to exploit, they are given access by the rulers who in turn get protection from the West and such leaders can squander the wealth of their countries as they so wish.

The report done by '60 Minutes Australia' examined America's relationship with Equatorial Guinea, titled "Dirty deeds" a relationship which doesn't care about abuses or the general well-being of the indigenous people. 60 minutes Australia revealed that Rock oil Australia and US giants like Exxon Mobil drill in the tiny nation of Equatorial Guinea. It is revealed in the report that just 12% of oil revenue goes to Equatorial Guinea according to Alex Vines, an African specialist of the Royal Institute of international affairs.

Steven Hayes runs America's corporate council on Africa which lobbies for American oil. He was questioned by the '60 minutes' journalist which some of their discussions follow thus:

Journalist: "Is there any depth to which you will not sink? Will you continue business with this man Obiang no matter how corrupt he is?"

Hayes: "I don't agree with it that American companies aren't following the fair-trade practices"

Journalist: Why don't you do the right thing, get out of Equatorial Guinea and turn your back on this corrupt dictator?

Hayes: Let's assume that every US oil company gets out of this country, how fast do you think it will be before the French and other international oil companies run there? A second, he added. So, what changes in Equatorial Guinea? Nothing.

By associating with this murderous tyrant, we in the West are party to a disgrace and contributing to a disaster in the making, the 60 seconds reporter said.

The riggs national bank, Obiang, like the Saudis, does business with Riggs where he controls 60 accounts in it including one in which oil companies deposit money in. In some of the accounts, the balances run for up to hundreds of millions of dollars. The presidential family owns a number of mansions in the outskirts of Washington DC, while his son, Teodoro Nguema Obiang who is the vice president owns his in Los Angeles, referred to as an 'African Prince' by the guard on duty at the time of the report by 60 minutes.

Predators will go to any length to keep their much coveted catch. Any foolish move that they make, their rival, especially the scavengers like the hyenas, wild dogs or vultures will jump in and take over the prey in less than no time. Reason why America will turn a blind eye to a corrupt and tyrannical government as that of Equatorial Guinea; open up her financial institutions like the Riggs bank where they store in their loots. In this manner, the rulers of such nations remain loyal to the predators or masters, if the corrupt rulers ever complain, their assets will be held by the big cats on grounds of human rights, something which is never mentioned so long as the rulers from the dark continent allow them to exploit the resources.

Again, I watched on the news channels how millions of Americans took to the streets in protest to the killing of George Floyd on May 25, 2020. The number of people on the streets sent shock waves across America and the world. Organizations that had vehemently rejected sympathy for the killing of African Americans by the police were quick to come out of their denial and show empathy for the people who had long been complaining. "What if these people never took to the streets and denounced police brutality as they did?'' I asked myself as I watched what unfolded on the television. "None of these organizations would have taken the stance they just did" the answer goes through my mind like a quiet breeze on a hot day, making me smile with its refreshing feeling.

"What a time to be alive" I say as events around the world unfold on cable news. "Who would have thought such things would someday erupt like the wildfires in the dry California forests?" From Africa, to Europe, to America, statues of one-time slave merchants and slave owners pulled down with loud cheers from participants.

The man who committed one of the worst atrocities in Africa, King Leopold of Belgium whose actions in the Congo are nothing less of a genocide, his statue not worthy to continue standing in public view in Belgium due to his actions in the Congo.

During Congo's 60th independence celebrations, Belgium's King apologized for the crimes his country committed in the Congo during their more than half a century occupation of the country. In his words, "I want to express my deepest regrets for the wounds of the past, the pain of which is revived today by discriminations that are still too present in our societies," the king wrote in the letter sent to President Félix Tshisekedi of the Democratic Republic of Congo. The king would, he added, "continue to fight against all forms of racism." he wrote.

Apologies offered and statues of past slavery henchmen being pulled down around the world while imperialism continues to kill Africa is of no importance. Slave trade and slavery dealt a blow on Africa, and at the era of the height of this human evil, colonialism came with its own consequences on the people of the Black continent. Neo-colonialism with imperialism picked up from where colonialism ended. The West has continuously maintained their grip on Africa as we see memories of their past atrocities being pulled down. They will not stop if Africans do not man up against these policies that keep impoverishing them, it is on Africa to right the wrongs of what the West did and continues to do on them.

One time African Union ambassador to the United States Dr. Arikana Chihombori-quao gave a pathway for Africans to correct the wrongs of 1884 when the Western powers converged in Berlin, Germany to brainstorm on ways of stabbing Africa. When I listened to her talk so passionately about stopping the French in particular, I quickly asked myself "will the buffaloes ever reason alike such that they can discover their strength in working as a team?" Before I knew it I was in my imaginary world once more, I couldn't help it but there I was, on the couch and imagining things as I always do. The calls made by Dr. Arikana and some other prominent Afro-activists were answered by Afro-diasporans in their millions and non-Afro-descendants alike who sympathize with the African people.

I envisaged Afro-descendants and other races coming together in response to calls made by some prominent activists, as a slow body of moving water they occupied some major cities of the world. Diasporans come out en mass with flags representing all nations of Africa and some carrying placards with messages such as "FREE AFRICA," FRENCH POLICIES = SILENT MASSACRE OF AFRICANS," others carrying pictures of Africans washed ashore and others with African migrants packed on boats on the Mediterranean ocean and still some had pictures of those who died in the jungles in South America.

Celebrities of African descent championing the marches, as some have taken upon themselves to challenge especially the French with their policies in Africa, Akon along with other artists get to perform

"If only I were a spirit, I would have gone around and touched every African and every descendant of Africa to reason with me and realize how much we can alter the status quo and prepare an Africa that her children will not flee from but desire to live in and enjoy all that she has, an Africa that her children will benefit from the vanilla that comes from Madagascar, one that the gold and diamond from Ghana to South Africa will glitter on her children and pearls from the Gulf of Aden and East Africa as a whole will beautify her children and not some faraway people.

An Africa whose vast oil and natural deposits from Northern Africa to West Africa will be explored by the Africans themselves for their own betterment and not for a few who in turn sell out the people. Again, in my imagination the images of the buffaloes and the predators come to mind, the realities of what the imperialists do to Africa come to mind, the naivete of Africans sadly comes to mind too. Then the question "Africa Why?" fills my mind.

Why can we not reason alike?

Why do you shoot your brothers and sisters when you don't even make guns?

Why do you live in poverty when you are rich?

‘Africa why?’ words from the song penned by Richard Kings, a young African who like millions of youths in Africa has pondered what children of the ‘Black continent’ did to deserve the fate meted on them. Richard Kings “Africa why”[4] *Words by Richrad kings in his song entitled ‘Africa why,’ questions which we all have asked in our own different ways without answers*

CHAPTER 13

NOT ALL BUFFALOES ARE FOOLISH AFTER ALL

(Tribute to JJ Rawlings & JP Magufuli)

CHAPTER 13

NOT ALL BUFFALOES ARE FOOLISH AFTER ALL

(Tribute to JJ Rawlings & JP Magufuli)

I watched a helpless buffalo as it laid down, its legs swinging in the air struggling as it laid back down while the lone lion tried to suffocate it with its powerful teeth gripping the buffalo's airpipe. I cursed the foolishness of the remaining herd of buffaloes as they stood by watching their friend being killed before their very own eyes. Then I wished they could do something and help the friend in captivity.

As if it heard my wish, one of the buffaloes mustered some courage and advanced cautiously towards the fallen friend under the lion's fangs. A few other buffaloes walked closely behind it. The roar of the lion scared them to back up, but the courageous one turned right back as it was close to the lion. Probably out of fear, the lion went running into the bushes after the buffaloes had returned and were close to the lion and their friend in captivity. "Not all buffaloes are foolish after all", I said as their friend got saved because of the bravery of the buffalo.

I woke up on November 12, 2020 with news of the death of former Ghanaian President, Jerry John Rowlings (JJ as Ghanaians enjoyed calling him). At the command of my voice, Alexa, the smart speaker that brings me news updates, tuned to the BBC world service for news updates. The passing of the former president of Ghana was one major news event of the day coming from Africa.

We had come to love the likes of JJ, Robert Mugabe of Zimbabwe, Muammar Gadhafi of Libya amongst and of recent John Pombe Magufuli of Tanzania and a few other African leaders who stood up against Western imperialism and actually challenged their policies in Africa.

It had not been long that a voice message circulated on social media, calling on France to step aside from African politics, accusing France of continuously forcing a currency on her former colonies which rather holds them back than advances them. The voice circulating on social media was one that I have personally fallen in love with. I had listened to him talk as a young man and as well as in his older age. The commanding tone of JJ Rowlings needed no one to tell who it was whenever we listened to him on the radio. The accusations pointing out to what the French continue to do that isn't for Africa's interest were from none other than the former Ghanaian president, JJ Rawlings. It is such stern rhetoric that we wished our leaders could use against imperialists, accompanied by action.

JJ's story of how he rose to power, coupled with his legacy are what some of us spend our whole lives hoping and praying that such could take place in our own nations. News of his death was a blow not just to Ghanaians but to the rest of the African continent.

If the buffaloes and the rest of the animals that fall under their class (herbivores) used by the flesh-eating animals (carnivores) as food fail to stand up against their counterparts, it doesn't mean that they fail to understand that they are used as food. They are just unable to exploit their strength because of their differences.

In the same way, Africans are not too dumb to realize that the West's presence on the continent from long ago, and now China's presence are to rip the riches of the continent at the expense of the people. Africans are just too naive not to realize their potential thereby continuing to be the victims in this jungle. When JJ's voice echoed calling on France to end her exploitation of the continent, he was simply voicing out what more than a billion people feel but are unable to act for fear of what the carnivores or their agents from within will do.

Just like the tens of brave sons and daughters of Africa who dared to oppose Western imperialism and got killed or deposed or least got isolated, JJ's death was quickly concluded by some to be the handiwork of the imperialists given his recent outburst against the French who are working hard to counter a plan by the West African economic blog, ECOWAS, which intends to come up with a single currency, the Eco. Should the West African nations which a good number of former French colonies (Ivory Coast, Mali, Benin, Senegal, Burkina Faso, Guinea, Guinea Bissau, Niger and Togo) belong, come up with a common currency, it will mean a change in monetary policies between France and her former colonies. JJ Rawlings did not compromise in calling out the French who have worked so hard to kill the plan which the Economic block of West African States has been trying to put in place.

Just as JJ Rawlings, John Pombe Joseph Magufuli caught most Africans' admiration with his simple outings as president. We viewed him as one who took himself low and closer to the people he presided over, their servant. The height of it was the news of a deal he canceled, one which his predecessor had signed with the Chinese.

JP Magufuli described the deal as exploitative and equally cited corruption and embezzlement by the ruling elites in Dar es Salaam. With such courage to take such decisions, the revelations got him lots of acclamations.

As I salute the efforts put up by JJ and JP, the same goes to the selfless patriots who, like the buffaloes that refused to think like their peers, risked their lives to liberate their type.

References

Abbot, F. ed. The Vatican II Council, Collegeville MN: The Liturgical Press, (1979)

Achermann, R. The Diversity of Interpretation of Secular Challenges, Kent: Osward Publication (2003)

Adams, Smith (2014) "The UN Rwanda and the Genocide"

Andrew Thomson, Searching for social Values for Reconstruction of Africa, Edinburgh: T. &T. Clark, 2006,2006, 76-77.

Freeman, D. H. A Philosophical study of Religion, New Jersey: The Craig Press, (1964)

Gerald Maxwell Thambudzai, Philosophy of African Traditions and search for the Common Good, San Francisco: Ignatius Press, 1996,45

Gregory Nagy, The Best of the Archaeans: Concept of the Hero, London: John's Hopkins University Press, (1979), 296-297

Ilo Stan, C. The Face of Africa Looking Beyond the Shadows, Ibadan: Spectrum Press, (2008)

John Paul II, Ecclesia in Africa (The Church in Africa) AAS 88, (1996)

Welch, C

Russell L. 2014, Evil: A Philosophical Investigation, Oxford University Press.

INDEX

A

Africa, 15,16,17,18,20,21,22,23,26,28, 31,35,36,37,40,41,42,43,44,45,46,49,50, 52,54,56,57,58,59,61,63,66,68,69,70,71, 73,74,75,77,79,80,81,82,83,84,85,88,92, 98,100,101,102,105,110,111,112,113,136 137,142,145,146,147,148,149,150,155,156 157,158,160,162,163,164,165,167,169,170

Algeria, 42,55,98,103,139,174

America, 35, 97, 104, 111,112,115,117,118, 143,154,159,163,165,166,168,176,178,180

B

Benin Empire, 21,105,106

Biafra, 148

British, 20,21,22,53, 60, 83,85,91,98,114, 154, 175

Bukina-faso, 58,147,173

Burundi, 139

C

Calabar, 21

Cameroon,16,22,29,30,34,35,45,68,69,70 72,75,93,104,112,165,166,169

Civil War, 53,148

Colonialist, 23,26,28,41,52,55,61,98,100, 130,136,137

Community, 19,42,47,110, 149,153,171,193

Congo-Brazzaville, 138

D

Deception,39,52,38

DR. Congo, 22, 23,26, 27, 28,33,34,35,41, 54.58,63,131,142,174,179

E

Egypt,79,80,82,105,124,131,138,172

Equatorial Guinea, 35, 176,177,178

Ethiopia, 79,80,82,105,124,126,138,149,158

European, 15,16,21,22,23,24,26,29,36,40 41,49,60,83,84,97,98,99,100,103,140,141, 145,146,147,149,164

Exploitation, 14,17,29,33,35,36,40, 57,63,84,124,142,152,160,174,186

F

France, 16, 42,43,44,45, 49, 97,111, 112, 136,137,139,142,146,147,148,156,157,164, 171,172,173,174,185

G

Gambia, 130,149,174

Genocide, 22,179

Germany,16,23,24,63,73,87,113,146,149,180

Ghana, 41,58,71,72,84,113,125,126,127,128, 149,150,157,165,166,167,169,174,181,185,186

Guinea-Bissau, 71,139

H

Hinterlands,17,18,20,75

History, 39,40,43,57,68,79,89,96,97, 99,100,103,104,105,111,116,122,123 124,126,151,152,160,170,191

Holland,16,46

Homeland, 148,166

I

In-Land, 18

Islands, 157

Ivory Coast, 128,137,147,150,173,174,186

K

Kenya, 85, 98, 149

L

Land,19,20,22,29,30,32,36,75,85, 103,117,118,121,122,126,128,129, 131,132,133,134,141,146,148,151, 158,160

Landscape, 130,155,172

*Legacy, 120,121,123,124,128,130
186*

Libya, 130,137,139,140,141,148,156

M

*Motherland, 111,114,118,157,165,
166,167,168*

Midlands, 99

N

*Nigeria, 21,72,93,105,113,114,115,121,
124,125,130,131,134,160,169,174*

P

Poland, 137

R

*Religion, 78,79,80,81,82,85,86,87,
88,89,90,91,92,93*

Resistance, 20, 54,98,151

Russia, 48, 97, 127, 146,173

Rwanda, 139,174,175,180,181,185,186

S

Second World War, 42,49,55

Senegal, 71,73,103,104,157,170,174,187

Scavengers, 48,65, 67,75,77,145

Scotland, 91

Silent Massacre,180

Somalia,138

Story, 51,52,63,85,99,103,104,113, 122,123,186

Sudan, 92,105,124

Switzerland, 45,130,173

T

Tanzania, 58,74,76,149,185

Togo, 147,148,150,174,187

W

War, 54,119,148,156

*West Africa, 22,23,24,31,45,68,69
70,71,111,124,125,126,128,130,165
173,181,186,187*

World war, 42,49

Z

Zanziba,138

Zimbabwe, 100,105,118,130,131,151,185

ABOUT THE BOOK

Africa is a magical, wondrous, and intriguing land but it is also harsh and unforgiving, why?

This book is a labor of love and a vital link to the richness and diversity of African struggles, *"Africa Why?"* honors the past, reminding us where we have been, and gives flight to our hopes for what is yet to come.

In a book as splendid in its wealth of information as it is breathtaking in scope, Teneng Muluh brings to light Africa's history and evolution, the majestic array of its landforms and environments, struggles missed with hope, love diluted with pain, the rich diversity of its people and the way of life for Africa. A book widely praised for its engaging and readable style, and is unrivalled in scope, both geographically and chronologically.

This book is one of the most alluring and easily understood dreams of our time. The tightly focused text moves quickly without sacrificing impact. It is a fascinating retelling of modern history that restores Africa to its rightful place. Creating a historical narrative that begins with the commencement of ugly truth about Africa.

"Africa Why?" is an intimate quest to explore the truth about people and the lost beauty of the environment. Elvis Muluh looks deeper inside his soul and rediscovers his

priorities through writing. He tries to paint the reason why Africa is often called the Cradle of Humankind.

THE AUTHOR

Teneng Muluh was born in Bamenda Cameroon in 1985. He started his primary school in Alahkuma, Bamenda before relocating to his village Buchi in the Pinyin clan as a result of financial hardship that the family faced.

After his A/L certification in 2007, he was faced with the challenges most African youths face, He eventually had the opportunity to leave home for the USA in 2015 where he now resides.

He is married with two children, a girl and boy while the third is on the way. His desire with this book is a rally call for the people of Africa and the descendants of the formally enslaved Africans all over the world.
His vision is to is to see Africa that would live up to its potentials as well as descendants of Africa the world at large discover the strength in their numbers.

Made in the USA
Monee, IL
04 November 2026

67381619R00108